Applied Mathematical Sciences
Volume 178

For further volumes:
http://www.springer.com/series/34

T. Insperger • G. Stépán

Semi-Discretization for Time-Delay Systems

Stability and Engineering Applications

 Springer

Tamás Insperger
Department of Applied Mechanics
Budapest University of Technology
and Economics
Müegyetem rkp. 5
1521 Budapest
Hungary
inspi@mm.bme.hu

Gábor Stépán
Department of Applied Mechanics
Budapest University of Technology
and Economics
Müegyetem rkp. 5
1521 Budapest
Hungary
stepan@mm.bme.hu

ISSN 0066-5452
ISBN 978-1-4614-3013-1   ISBN 978-1-4614-0335-7 (eBook)
DOI 10.1007/978-1-4614-0335-7
Springer New York Dordrecht Heidelberg London

Mathematical Subject Classification (2010): 34K06, 34K13, 34K20, 34K28, 34K35, 37N15, 37N35

Springer is part of Springer Science+Business Media (www.springer.com)

To our parents and to our Ágies

Preface

Time delay always arises in engineering models, where the rates of change of state variables depend both on present and on past state variables of the system. Control processes with feedback delay, regenerative machine tool chatter, wheel shimmy models including the elastic contact between the tire and the road, car-following traffic models with the reaction time of the drivers, human motion control with reflex delay, can be mentioned as examples. The analysis of these systems requires the characterization of their local behavior around a desired position or a desired (possibly periodic) path. Such properties can be described by stability charts that present the stability of the linearized system in the plane of the system parameters. These stability charts provide a useful tool for engineers, since they present an overview on the effects of system parameters on the local dynamics of the system.

The main differences between systems with and without time delay is that time delay produces an infinite-dimensional dynamics as opposed to the finite-dimensional dynamics of delay-free systems. For simple time-delay systems, stability charts can be derived analytically. However for complex systems, for instance, when the time-delay effect is coupled with parametric excitation, only numerical techniques can be used.

The scope of this book is to present a numerical technique, called the semi-discretization method, for the stability analysis of linear time-periodic time-delay systems, which is also an essential tool in the study of periodic motions of nonlinear time-delay systems. Semi-discretization is a well-known technique used, for example, in the finite element analysis of solid bodies, or in computational fluid mechanics, where the corresponding partial differential equations are discretized along the spatial coordinates only, while the time coordinates are unchanged. In case of time-delay systems, semi-discretization results in the discretization of delayed terms only, while the actual-time-domain terms are not discretized. In this way, the infinite-dimensional system is approximated by a finite-dimensional one.

The structure of the book is as follows. Chapter 1 gives some introduction to linear time-delay systems. Chapter 2 deals with the construction of the stability charts for some fundamental delay-differential equations. The semi-discretization method is presented in Chapter 3 including higher-order methods, rate convergence estimates, and numerical issues. The semi-discretization method is applied in Chapter 4 to some Newtonian examples with different delay types, such as single point delay, multiple delays, distributed delay, and time-periodic delay. Finally, Chapter 5 presents real-world mechanical engineering applications. Turning and milling processes are considered with varying spindle speed, resulting in time-periodic time delays. Then, the so-called act-and-wait control concept is introduced, and it is analyzed through applications to the stick-balancing problem and to a force-control process with feedback delay. It is shown that the inclusion of waiting periods in the control rule may have a stabilizing effect. This provides the surprising conclusion that doing nothing and rather waiting for the response of a previous action might be a superior control strategy for systems with feedback delay. Finally, the stick-balancing model with reflex delay is investigated in the case of parametric forcing at the stick's base. The book concludes with an appendix that contains Matlab codes for the semi-discretization of the examples presented in Chapter 4.

The book is designed for graduate and PhD students as well as for researchers working in the fields of mechanical, electrical, and chemical engineering, control theory, biomechanics, population dynamics, neurophysiology, even climate research in which time-delay models occur.

The book is based on the authors' research work over the last 10 years, but many colleagues have contributed to different parts. Hereby, the authors thank and acknowledge the useful discussions with and comments of Mikel Zatarain, Jokin Muñoa, Grégoire Peigné, and Sébastien Seguy regarding the computational efficiency of the semi-discretization for different milling applications. The helpful consultations and joint works related to the mathematical issues of the method with Janos Turi, Ferenc Hartung, and Barnabás Garay are greatly appreciated. The comments and novel ideas provided by our young colleagues Zoltán Dombóvári and Dániel Bachrathy are gratefully recognized. Finally, the inspiring long-term cooperation and mutual comparative studies for machining operations with Philip V. Bayly, Brian P. Mann, and Firas A. Khasawneh are gratefully acknowledged.

The appearance of this book and the related research work during recent years were supported by the Hungarian National Science Foundation (OTKA) under grant no. K72911 and K68910 and the János Bolyai Research Scholarship of the Hungarian Academy of Sciences. This work is linked to the scientific program of the "Development of quality-oriented and harmonized R+D+I strategy and functional model at BME" project. This project is supported by the New Hungary Development Plan (Project ID: TÁMOP-4.2.1/B-09/1/KMR-2010-0002). This book is related to activities performed within the DYNXPERTS project, funded by the European Commission FP7 Factories of the Future with Grant Number 260073.

Budapest,   *Tamás Insperger*
April 2011   *Gábor Stépán*

Contents

Chapter 1
Introducing Delay in Linear Time-Periodic Systems

Dynamical systems have been described with differential equations since the appearance of the differential calculus; Newton's second law could be considered one of the first examples. A differential equation can serve as a model for how the rate of change of state depends on the present state of a system. However, the rate of change of state may depend on past states, too. It has been known for a long time that several problems can be described by models including past effects. One of the classical examples is the predator–prey model of Volterra [288], where the growth rate of predators depends not only on the present quality of food (say, prey), but also on past quantities (in the period of gestation, say). The first delay models in engineering appeared for wheel shimmy [230] and for ship stabilization [194] in the early 1940s. There are several other engineering applications in which time delay plays a crucial role. As recognized in the late 1940s with the development of control theory, time delay typically arises in feedback control systems due to the finite speed of information transmission and data processing [284, 252]. Another typical application is the stability of machining processes, where time delay appears due to the surface regeneration by the cutting edge [280, 281, 256, 5]. Similar equations describe the car-following traffic models involving the reaction time of drivers [213, 214, 215]. Reflex delay is also a relevant issue to human motion control [24, 259, 192, 13]. Time delay also plays important role in population dynamics [160, 251], in neural networks [49, 216], and in epidemiology models [226, 2].

Systems whose rate of change of state depends on states at deviating arguments are generally described by functional differential equations (FDEs). According to Myshkis [203], FDEs are equations involving the function $x(t)$ of one scalar argument t (called time) and its derivatives for several values of argument t. FDEs can be categorized into retarded, neutral, and advanced types (see, e.g., [74, 152]). If the rate of change of state depends on past states of the system, then the corresponding mathematical model is a retarded functional differential equation (RFDE). If the rate of change of state depends on its own past values as well, then the corresponding equation is called a neutral functional differential equation (NFDE). If the rate of change of state depends on past values of higher derivatives of the state, then the system is described by an advanced functional differential equation (AFDE). These equations are also referred to as FDEs of retarded, neutral, or advanced type. While RFDEs and NFDEs have many practical applications, AFDEs are rarely used in engineering modeling due to their inverted causality. Note, however, that there are some special problems even in Newtonian mechanics where the governing equations are related to AFDEs [136, 137].

The literature on FDEs is quite extensive. Several books have appeared summarizing the most important theorems; see, for instance, the books by Myshkis [203], Bellman and Cooke [27], Èl'sgol'c [74], Halanay [98], Hale [99], Driver [72], Kolmanovskii and Nosov [153], Hale and Lunel [100], Kolmanovskii and Myshkis [152], Diekmann et al. [64], just to mention a few. There are also several books dealing with different applications and numerical techniques; see for instance, Stepan [255], Kuang [160], Kuang and Cong [159], Niculescu [207], Hu and Wang [113], Bellen and Zennaro [26], Gu et al. [91], Zhong [306], Michiels and Niculescu [187], Kushner [163], Erneux [79], Balachandran et al. [20], Lakshmanan and Senthilkumar [165], Smith [251], and Yi et al. [299]. It is known that discretization techniques preserve asymptotic stability for RFDEs (see, e.g., [95] or [85]); however, this is not true for NFDEs and AFDEs in general (see, e.g., [81] and [136], respectively).

RFDE is a mathematical terminology. In the engineering literature, RFDEs are referred to as delay-differential equations (DDEs), or simply delay equations. In this monograph, we follow the latter terminology, and use the term DDE rather than RFDE.

This monograph deals with the stability analysis of *linear time-periodic DDEs* using the *semi-discretization method*. These equations often arise during the analysis of delayed systems, since the stability properties of the periodic orbits of nonlinear DDEs are described by linear time-periodic DDEs [143, 158]. This introductory chapter gives a brief overview on some special cases of linear DDEs. The corresponding basic theory is essential for constructing the analytical examples of Chapter 2, which then serve as references for the tests of the numerical method introduced in Chapter 3. The last two chapters investigate pure Newtonian examples of delayed oscillators and the dynamics of real-world engineering problems modeled by time-periodic DDEs.

1.1 Linear Autonomous ODEs

Linear autonomous ordinary differential equations (ODEs) have the general form

$$\dot{\mathbf{x}}(t) = \mathbf{A}\mathbf{x}(t) , \qquad (1.1)$$

where $\mathbf{x}(t) \in \mathbb{R}^n$, $\mathbf{A}$ is an $n \times n$ matrix, and

$$\dot{\mathbf{x}} = \frac{d\mathbf{x}}{dt} = \mathrm{col}\left(\frac{dx_1}{dt} \ \ \frac{dx_2}{dt} \ \ \cdots \ \ \frac{dx_n}{dt} \right)$$

with $x_1, x_2, \ldots, x_n$ being the elements of vector $\mathbf{x}$. For a given initial value $\mathbf{x}(0)$, the solution of (1.1) can be written in the form

$$\mathbf{x}(t) = e^{\mathbf{A}t}\mathbf{x}(0) \,, \tag{1.2}$$

where $e^{\mathbf{A}t}$ is the exponential of matrix $\mathbf{A}t$, defined by the Taylor series of the exponential function (see Appendix A.1). For a general overview on matrix exponentials, see the book of Hirsch and Smale [108], or the book of Perko [219].

The stability of the trivial solution $\mathbf{x}(t) \equiv \mathbf{0}$ is determined by the eigenvalues λ_j, $j = 1, 2, \ldots, n$, of the coefficient matrix $\mathbf{A}$. These eigenvalues are the *characteristic exponents* of (1.1), but they are often called *characteristic roots* or *poles*, too. If each λ_j is unique in the minimal polynomial of $\mathbf{A}$, then each solution of (1.1) can be written in the form

$$\mathbf{x}(t) = \sum_{j=1}^{n} \mathbf{C}_j e^{\lambda_j t} \,, \tag{1.3}$$

with $\mathbf{C}_j \in \mathbb{C}^n$ being appropriate vectors depending on the initial condition. If the characteristic exponents have negative real parts, i.e., $\mathrm{Re}\,\lambda_j < 0$ for all $j = 1, 2, \ldots, n$, then the trivial solution of (1.1) is asymptotically stable. In the general case, the characteristic exponents can be determined by solving the characteristic equation

$$\det(\lambda\mathbf{I} - \mathbf{A}) = 0 \,, \tag{1.4}$$

where $\mathbf{I}$ stands for the $n \times n$ identity matrix. Development of (1.4) results in an nth-degree polynomial of λ, whose roots (i.e., the characteristic exponents) can be determined by a number of numerical methods. Stability analysis, however, does not require the exact calculation of the characteristic exponents; only the sign of the real part of the critical (i.e., rightmost) exponent must be determined. This analysis can be performed by the celebrated Routh–Hurwitz criterion [227, 114], which gives a necessary and sufficient condition for stability based on the coefficients of the characteristic polynomial (for details, see Appendix A.2).

Depending on the location of the critical characteristic exponents, there are two typical mechanisms for loss of stability of linear autonomous systems [92]:

1. The critical characteristic exponents form a complex conjugate pair moving from the left-hand side of the complex plane to the right-hand side; they cross the imaginary axis, as shown by case (a) in Figure 1.1. This case is an essential necessary condition for the so-called *Hopf* (or *Andronov–Hopf* or *Poincaré–Andronov–Hopf*) bifurcation of the corresponding nonlinear system, for which the equation under analysis is the variational system. The systematic study of the conditions and a proof of the corresponding bifurcation theorem have been done by Andronov and Leontovich [10] for the two-dimensional case, and by Hopf [109] for the n-dimensional case. According to the theory of nonlinear systems, either stable or unstable periodic motion may exist around the equilibrium of the corresponding nonlinear system, called supercritical and subcritical bifurcation, respectively.

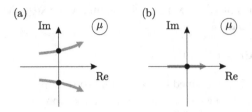

Fig. 1.1 Critical characteristic exponents for linear
autonomous ODEs: (a) Hopf
bifurcation, (b) and saddle-
node bifurcation.

2. The critical characteristic exponent is a real one moving from the left-hand side
   of the complex plane to the right-hand side through the origin, as shown by case
   (b) in Figure 1.1. This case is called *saddle-node* bifurcation of the corresponding
   nonlinear system.

1.2 Linear Periodic ODEs

The general form of linear periodic ODEs reads

$$\dot{\mathbf{x}}(t) = \mathbf{A}(t)\mathbf{x}(t) , \quad \mathbf{A}(t) = \mathbf{A}(t + T) , \tag{1.5}$$

with $\mathbf{x}(t) \in \mathbb{R}^n$. Here, the $n \times n$ coefficient matrix $\mathbf{A}(t)$ is time-periodic at period
T, called the *principal period* in contrast to the constant-coefficient matrix of the
autonomous system (1.1). The main theorems on general periodic systems are sum-
marized in the book of Farkas [83].

For periodic ODEs, a stability condition is provided by the Floquet theory [84].
The solution of (1.5) with the initial condition $\mathbf{x}(0)$ is given by $\mathbf{x}(t) = \mathbf{\Phi}(t)\mathbf{x}(0)$,
where $\mathbf{\Phi}(t)$ is a fundamental matrix of (1.5). According to the Floquet theory, the
fundamental matrix can be written in the form $\mathbf{\Phi}(t) = \mathbf{P}(t)\mathrm{e}^{\mathbf{B}t}$, where $\mathbf{P}(t) = \mathbf{P}(t + T)$
is a periodic matrix with initial value $\mathbf{P}(0) = \mathbf{I}$, and $\mathbf{B}$ is a constant matrix. The
matrix $\mathbf{\Phi}(T) = \mathrm{e}^{\mathbf{B}T}$ is called the *monodromy matrix* (or *principal matrix* or *Floquet
transition matrix*) of (1.5). This matrix gives the connection between the initial state
and the state one principal period later: $\mathbf{x}(T) = \mathbf{\Phi}(T)\mathbf{x}(0)$.

The eigenvalues of $\mathbf{\Phi}(T)$ are the *characteristic multipliers* (μ_j, $j = 1, 2, \ldots, n$)
(also called *Floquet multipliers* or the *poles* of $\mathbf{\Phi}(T)$) calculated from

$$\det(\mu\mathbf{I} - \mathbf{\Phi}(T)) = 0 . \tag{1.6}$$

The eigenvalues of matrix $\mathbf{B}$ are the *characteristic exponents* (λ_j, $j = 1, 2, \ldots, n$)
given by

$$\det(\lambda\mathbf{I} - \mathbf{B}) = 0 . \tag{1.7}$$

If μ is a characteristic multiplier, then there are characteristic exponents λ such that
$\mu = \exp(\lambda T)$, and vice versa. Due to the periodicity of the complex exponential
function, each characteristic multiplier is associated with infinitely many character-

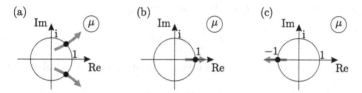

Fig. 1.2 Critical characteristic multipliers for periodic systems: (a) secondary Hopf bifurcation, (b) cyclic-fold bifurcation, and (c) period-doubling bifurcation.

istic exponents of the form $\lambda_k = \gamma + i(\omega + k2\pi/T)$, where $\gamma, \omega \in \mathbb{R}$, $k \in \mathbb{Z}$, and $T\omega \in (-\pi, \pi]$.

The trivial solution $\mathbf{x}(t) \equiv \mathbf{0}$ of (1.5) is asymptotically stable if and only if all the characteristic multipliers have modulus less than one, that is, all the characteristic exponents have negative real parts.

Similarly to autonomous systems, the basic types of loss of stability can be classified according to the location of the critical characteristic multipliers [92]. For periodic systems, there are three typical cases:

1. The critical characteristic multipliers form a complex conjugate pair crossing the unit circle, i.e., $|\mu| = 1$ and $|\bar{\mu}| = 1$, as shown by case (a) in Figure 1.2. This case is topologically equivalent to the Hopf bifurcation of autonomous systems and is called *secondary Hopf* (or *Neimark–Sacker*) bifurcation.
2. The critical characteristic multiplier is real and crosses the unit circle at $+1$, as shown by case (b) in Figure 1.2. The bifurcation that arises is topologically equivalent to the saddle-node bifurcation of autonomous systems and is called *cyclic-fold* (or *period-one*) bifurcation.
3. The critical characteristic multiplier is real and crosses the unit circle at -1, as shown by case (c) in Figure 1.2. There is no topologically equivalent type of bifurcation for autonomous systems. This case is called *period-doubling* (or *period-two* or *flip*) bifurcation.

Generally, the monodromy matrix cannot be determined in closed form, but there exist several numerical and semi-analytical techniques to approximate it, such as Hill's infinite determinant method and its generalizations [107, 266, 32, 205], the method of strained parameters [205], the method of multiple scales [205], and the Chebyshev polynomial approach [247, 246]. A simple numerical method is the piecewise constant approximation of the periodic matrix $\mathbf{A}(t)$ in the form

$$\mathbf{A}(t) \approx \mathbf{A}_i := \int_{(i-1)h}^{ih} \mathbf{A}(s)\,\mathrm{d}s \,, \quad t \in [t_i, t_{i+1}) \,, \tag{1.8}$$

where $t_i = ih$ is the discrete time with $i \in \mathbb{Z}$, $h = T/p$ is the length of the discretization step, and p is an integer [111, 83]. The original system can be approximated by

$$\dot{\mathbf{y}}(t) = \mathbf{A}_i \mathbf{y}(t) \,, \quad t \in [t_i, t_{i+1}) \,, \tag{1.9}$$

for which the solution over a discretization interval is

$$\mathbf{y}(t_{i+1}) = e^{\mathbf{A}_i h} \mathbf{y}(t_i) . \tag{1.10}$$

Application of (1.10) over p repeated discretization steps with initial state $\mathbf{y}(0)$ results in

$$\mathbf{y}(T) = \tilde{\boldsymbol{\Phi}}(T)\mathbf{y}(0) , \tag{1.11}$$

where

$$\tilde{\boldsymbol{\Phi}}(T) = e^{\mathbf{A}_{p-1} h} e^{\mathbf{A}_{p-2} h} \cdots e^{\mathbf{A}_0 h} \tag{1.12}$$

is an approximation for the monodromy matrix $\boldsymbol{\Phi}(T)$. Eigenvalue analysis of $\tilde{\boldsymbol{\Phi}}(T)$ gives then an approximate description of the stability properties of (1.5). A higher-order generalization of this piecewise constant approximation technique is the method of Magnus expansion, which involves higher-order terms of the so-called Magnus series of the logarithm of the fundamental matrix $\boldsymbol{\Phi}(h)$ (see, e.g., [175, 138, 139, 46]). Approximation (1.8) corresponds to the first-order Magnus expansion of $\ln(\boldsymbol{\Phi}(h))$.

1.3 Linear Autonomous DDEs

The general form of linear autonomous DDEs is

$$\dot{\mathbf{x}}(t) = \mathbf{L}(\mathbf{x}_t) , \tag{1.13}$$

where $\mathbf{L} : C \to \mathbb{R}^n$ is a continuous linear functional (C is the Banach space of continuous functions) and the continuous function $\mathbf{x}_t$ is defined by the shift

$$\mathbf{x}_t(\vartheta) = \mathbf{x}(t + \vartheta) , \quad \vartheta \in [-\sigma, 0] . \tag{1.14}$$

According to the Riesz representation theorem (see [99]), the linear functional $\mathbf{L}$ can be represented in the matrix form

$$\mathbf{L}(\mathbf{x}_t) = \int_{-\sigma}^0 d\boldsymbol{\eta}(\vartheta) \, \mathbf{x}(t + \vartheta) , \tag{1.15}$$

where $\boldsymbol{\eta} : [-\sigma, 0] \to \mathbb{R}^{n \times n}$ is a matrix function of bounded variation, and the integral is a Stieltjes one, i.e., (1.15) contains both point delays and distributed delays.

The characteristic equation can be obtained by substituting the nontrivial solution $\mathbf{x}(t) = \mathbf{C} e^{\lambda t}$, $\mathbf{C} \in \mathbb{C}^n$, into (1.13), which gives

$$\underbrace{\det \left(\lambda \mathbf{I} - \int_{-\sigma}^0 e^{\lambda \vartheta} \, d\boldsymbol{\eta}(\vartheta) \right)}_{:= D(\lambda)} = 0 . \tag{1.16}$$

The left-hand side of this equation defines the characteristic function $D(\lambda)$ of (1.13). The characteristic exponents are the zeros of the characteristic function. As opposed to the characteristic polynomial of autonomous ODEs, the characteristic function $D(\lambda)$ has, in general, an infinite number of zeros in the complex plane, all of which should be considered during the stability analysis. Stability charts that present the stability properties as a function of the system parameters have therefore a rich and intricate structure even for the simplest DDEs.

DDEs containing only point/discrete delays can be given in the form

$$\dot{\mathbf{x}}(t) = \mathbf{A}\mathbf{x}(t) + \sum_{j=1}^{g} \mathbf{B}_j \mathbf{x}(t - \tau_j) , \qquad (1.17)$$

where $\mathbf{A}$ and the $\mathbf{B}_j$'s are $n \times n$ matrices, $\tau_j > 0$ for all j, and $g \in \mathbb{Z}^+$. In this case, only discrete values of the past have influence on the present rate of change of state.

An example of a DDE with distributed delay is

$$\dot{\mathbf{x}}(t) = \mathbf{A}\mathbf{x}(t) + \int_{-\sigma_1}^{-\sigma_2} \mathbf{K}(\vartheta)\, \mathbf{x}(t + \vartheta)\, d\vartheta , \qquad (1.18)$$

where $\mathbf{K}(\vartheta)$ is an $n \times n$ measurable kernel function, $\sigma_1, \sigma_2 \in \mathbb{R}$, and $\sigma_1 > \sigma_2 \geq 0$. The kernel function $\mathbf{K}(\vartheta)$ describes the weight of the past effects over the interval $[t - \sigma_1, t - \sigma_2]$. If the kernel is a constant matrix multiplied by the shifted Dirac delta distribution, i.e., $\mathbf{K}(\vartheta) = \mathbf{K}_0\, \delta(\vartheta + \tau)$ with $\sigma_1 \leq \tau \leq \sigma_2$, then the integral in (1.18) gives the point delay $\mathbf{K}_0\, \mathbf{x}(t - \tau)$.

Linear autonomous DDEs with distributed delay and with a finite number of point delays can be given in the general form

$$\dot{\mathbf{x}}(t) = \int_{-\sigma}^{0} \mathbf{K}(\vartheta)\mathbf{x}(t + \vartheta)\, d\vartheta , \qquad (1.19)$$

where $\mathbf{K}(\vartheta)$ is an $n \times n$ measurable kernel function that may comprise a measurable distribution and finitely many shifted Dirac delta distributions. That is, $\mathbf{K}(\vartheta)$ can also be given in the form

$$\mathbf{K}(\vartheta) = \mathbf{W}(\vartheta) + \sum_{j=1}^{g} \mathbf{B}_j \delta(\vartheta + \tau_j) , \qquad (1.20)$$

where $\mathbf{W}(\vartheta)$ is an $n \times n$ measurable function (a weight function), the $\mathbf{B}_j$'s are $n \times n$ constant matrices, $\delta(\vartheta)$ denotes the Dirac delta distribution, $\tau_j \geq 0$ for all j, and $g \in \mathbb{N}$. Thus, (1.19) can be written as

$$\dot{\mathbf{x}}(t) = \int_{-\sigma}^{0} \mathbf{W}(\vartheta)\mathbf{x}(t + \vartheta)d\vartheta + \sum_{j=1}^{g} \mathbf{B}_j \mathbf{x}(t - \tau_j) . \qquad (1.21)$$

A necessary and sufficient condition for the asymptotic stability of DDE (1.13) with (1.15) is that all the infinite number of characteristic exponents have negative real parts and there exist a scalar $\nu > 0$ such that

$$\int_{-\infty}^{0} e^{-\nu\vartheta} \left| d\eta_{jk}(\vartheta) \right| < \infty , \quad j,k = 1, 2, \ldots, n , \tag{1.22}$$

where $\eta_{jk}(\vartheta)$ are the elements of $\boldsymbol{\eta}(\vartheta)$. Condition (1.22) means that the past effect decays exponentially in the past. Obviously, this condition holds if σ in the lower limit of the integral in (1.15) is finite.

Although there are infinitely many characteristic exponents, it is not necessary to compute all of them, since stability analysis requires only the sign of the real part of the rightmost one(s). There exist several analytical and semi-analytical methods to derive the stability conditions for the system parameters. The first attempts for determining stability criteria for first- and second-order scalar DDEs were made by Bellmann and Cooke [27] and by Bhatt and Hsu [28]. They used the D-subdivision method of Neimark [206] combined with a theorem of Pontryagin [221]. The book of Kolmanovskii and Nosov [153] summarizes the main theorems on the stability of DDEs, and contains several examples as well. A sophisticated method was developed by Stepan [255] (generalized also by Hassard [103]) that can be applied even for a combination of multiple point delays and for distributed delays. There exist several efficient numerical methods to determine the rightmost exponents for a delayed system; see, for instance, the celebrated DDE-BIFTOOL developed by Engelborghs et al. [76, 77], the pseudospectral differencing method by Breda et al. [34, 35], the cluster treatment method by Olgac and Sipahi [210, 211], the Galerkin projection by Wahi and Chatterjee [289, 290], the mapping algorithm by Vyhlídal and Zítek [287], the harmonic balance by Liu and Kalmár-Nagy [171], or the Lambert W function approach by Ulsoy et al. [14, 298].

The stability properties of DDEs are often represented in the form of stability charts that show the stable and unstable domains, or alternatively, the number of unstable characteristic exponents (also called instability degree) in the space of system parameters. Stability charts for autonomous DDEs can be constructed by the *D-subdivision method*. The curves where changes in the number of unstable exponents happen are given by the so-called *D-curves* (also called exponent-crossing curves or transition curves) given by

$$R(\omega) = 0 , \quad S(\omega) = 0 , \quad \omega \in [0, \infty) , \tag{1.23}$$

where

$$R(\omega) := \operatorname{Re} D(\mathrm{i}\omega) , \quad S(\omega) := \operatorname{Im} D(\mathrm{i}\omega) , \tag{1.24}$$

with $D(\lambda)$ being the characteristic function defined in (1.16) and ω the parameter of the curves [256]. Due to the continuity of the characteristic exponents with respect to changes in the system parameters (see, for instance, [187]), the D-curves separate the parameter space into domains where the numbers of unstable characteristic exponents are constant. The determination of these numbers for the individual domains is not a trivial task. One technique is to calculate the *exponent-crossing direction* (also called root-crossing direction or root tendency) along the D-curves, which is the sign of the partial derivative of the real part of the characteristic exponent with respect to one of the system parameters. If the number of unstable exponents is known for at least one point in one domain, then it can be determined for all the other domains by considering the exponent-crossing direction along the D-curves. The *stability boundaries* are the D-curves bounded the domains with zero unstable characteristic exponent.

Alternatively, Stepan's formulas [255] can also be used to determine the number of unstable characteristic exponents in a simple and elegant way. This technique requires the analysis of the functions $R(\omega)$ and $S(\omega)$ defined in (1.24) only, without the analysis of the exponent-crossing direction. Assume that the characteristic function $D(\lambda)$ associated with (1.13) has no zeros on the imaginary axis and (1.22) holds. If the dimension n of (1.13) is even, i.e., $n = 2m$ with m being an integer, then the number of unstable exponents is

$$N = m + (-1)^m \sum_{k=1}^{r} (-1)^{k+1} \text{sgn}\, S(\rho_k) , \qquad (1.25)$$

where $\rho_1 \geq \cdots \geq \rho_r > 0$ are the positive real zeros of $R(\omega)$. If the dimension n of (1.13) is odd, i.e., $n = 2m + 1$ with m being an integer, then the number of unstable exponents is

$$N = m + \frac{1}{2} + (-1)^m \left(\frac{1}{2}(-1)^s \text{sgn}\, R(0) + \sum_{k=1}^{s-1} (-1)^k \text{sgn}\, R(\sigma_k) \right) , \qquad (1.26)$$

where $\sigma_1 \geq \cdots \geq \sigma_s = 0$ are the nonnegative real zeros of $S(\omega)$. For further details and for an exact proof, see Theorems 2.15 and 2.16 in [255].

1.4 Linear Time-Periodic DDEs

Linear time-periodic DDEs have the general form

$$\dot{\mathbf{x}}(t) = \mathbf{L}(t, \mathbf{x}_t) , \quad \mathbf{L}(t + T) = \mathbf{L}(t) , \qquad (1.27)$$

where $\mathbf{x}_t$ is a continuous function defined by (1.14), $\mathbf{L} : \mathbb{R} \times C \to \mathbb{R}^n$ is continuous and linear in $\mathbf{x}_t$. According to the Riesz representation theorem, the functional $\mathbf{L}$ can be written in the Stieltjes integral form

$$\mathbf{L}(t, \mathbf{x}_t) = \int_{-\sigma}^{0} d_\vartheta \boldsymbol{\eta}(t, \vartheta) \mathbf{x}(t + \vartheta) , \qquad (1.28)$$

where $\boldsymbol{\eta} : \mathbb{R} \times [-\sigma, 0] \to \mathbb{R}^{n \times n}$ is a matrix function of bounded variation in $\vartheta \in [-\sigma, 0]$.

Linear time-periodic DDEs with constant point delays can be defined as

$$\dot{\mathbf{x}}(t) = \mathbf{A}(t)\mathbf{x}(t) + \sum_{j=1}^{g} \mathbf{B}_j(t)\,\mathbf{x}(t - \tau_j)\,, \tag{1.29}$$

$$\mathbf{A}(t + T) = \mathbf{A}(t)\,, \quad \mathbf{B}_j(t + T) = \mathbf{B}_j(t)\,,$$

where $\mathbf{A}(t)$ and the $\mathbf{B}_j(t)$'s are $n \times n$ matrices, $\tau_j > 0$ for all j, and $g \in \mathbb{Z}^+$. Linear time-periodic DDEs with distributed delay read

$$\dot{\mathbf{x}}(t) = \mathbf{A}(t)\mathbf{x}(t) + \int_{-\sigma_1}^{-\sigma_2} \mathbf{K}(\vartheta, t)\,\mathbf{x}(t + \vartheta)\,\mathrm{d}\vartheta\,, \tag{1.30}$$

$$\mathbf{A}(t + T) = \mathbf{A}(t)\,, \quad \mathbf{K}(\vartheta, t + T) = \mathbf{K}(\vartheta, t)\,,$$

where $\mathbf{K}(\vartheta, t)$ is the time-periodic $n \times n$ kernel function, $\sigma_1, \sigma_2 \in \mathbb{R}$, and $\sigma_1 > \sigma_2 \geq 0$. A special case of (1.30) occurs when the kernel function is a constant matrix multiplied by a time-periodic Dirac delta distribution, i.e., $\mathbf{K}(\vartheta, t) = \mathbf{K}_0 \delta(\vartheta - \tau(t))$ with $\sigma_1 \leq \tau(t) \leq \sigma_2$ and $\tau(t + T) = \tau(t)$. In this case, the integral in (1.30) gives the time-periodic point delay $\mathbf{K}_0 \mathbf{x}(t - \tau(t))$. The general form of linear time-periodic DDEs with time-periodic point delays is

$$\dot{\mathbf{x}}(t) = \mathbf{A}(t)\mathbf{x}(t) + \sum_{j=1}^{g} \mathbf{B}_j(t)\,\mathbf{x}(t - \tau_j(t))\,, \tag{1.31}$$

$$\mathbf{A}(t + T) = \mathbf{A}(t)\,, \quad \mathbf{B}_j(t + T) = \mathbf{B}_j(t)\,, \quad \tau_j(t + T) = \tau_j(t)\,.$$

Linear time-periodic DDEs with distributed delay and with a finite number of point delays can be given in the general form

$$\dot{\mathbf{x}}(t) = \int_{-\sigma}^{0} \mathbf{K}(\vartheta, t)\mathbf{x}(t + \vartheta)\,\mathrm{d}\vartheta\,, \tag{1.32}$$

where $\mathbf{K}(\vartheta, t)$ is an $n \times n$ measurable kernel function that can be written in the form

$$\mathbf{K}(\vartheta, t) = \mathbf{W}(\vartheta, t) + \sum_{j=1}^{g} \mathbf{B}_j(t)\delta(\vartheta + \tau_j(t))\,. \tag{1.33}$$

Here $\mathbf{W}(\vartheta, t + T) = \mathbf{W}(\vartheta, t)$ is a time-periodic $n \times n$ measurable function (a time-periodic weight function), $\mathbf{B}_j(t + T) = \mathbf{B}_j(t)$ are $n \times n$ time-periodic matrices, $\delta(\vartheta)$ denotes the Dirac delta distribution, $\tau_j(t) \geq 0$ for all j, and $g \in \mathbb{N}$. Thus, (1.32) can be written as

$$\dot{\mathbf{x}}(t) = \int_{-\sigma}^{0} \mathbf{W}(\vartheta, t)\mathbf{x}(t + \vartheta)\mathrm{d}\vartheta + \sum_{j=1}^{g} \mathbf{B}_j(t)\mathbf{x}(t - \tau_j(t))\,. \tag{1.34}$$

If the delay depends not only on the time but also on the state, then the corresponding equation is called a DDE with state-dependent delay. A simple example of such a DDE is

$$\dot{\mathbf{x}}(t) = \mathbf{A}\mathbf{x}(t) + \mathbf{B}\,\mathbf{x}(t - \tau(t, \mathbf{x}(t))) \qquad (1.35)$$

with $\tau(t, \mathbf{x}(t)) \geq 0$. State-dependent delays arose first in the two-body problem of classical electrodynamics [71], but they appear in many other fields, such as population models [208, 160, 4], automatic position control [292], neural networks models [21], and machine tool vibration theory [132]. A survey about DDEs with state-dependent delays is given in [102]. DDEs with state-dependent delays are always nonlinear, since the state appears in its own argument, and are therefore not analyzed in this monograph.

According to the Floquet theory of DDEs [97, 100], the solution segment $\mathbf{x}_t$ for (1.27) associated with the initial function $\mathbf{x}_0$ can be given as $\mathbf{x}_t = \mathcal{U}(t)\mathbf{x}_0$, where $\mathcal{U}(t)$ is the solution operator (infinitesimal generator). The stability of the system is determined by the spectrum of the corresponding monodromy operator $\mathcal{U}(T)$. The nonzero elements of the spectrum of $\mathcal{U}(T)$ are called *characteristic multipliers* (also referred to as *Floquet multipliers* or *poles*) and are defined by

$$\mathrm{Ker}(\mu \mathcal{I} - \mathcal{U}(T)) \setminus \{\mathbf{0}\} = \emptyset\,, \quad \mu \neq 0\,, \qquad (1.36)$$

or

$$\mathrm{Ker}(\mu \mathcal{I} - \mathcal{U}(T)) \neq \{\mathbf{0}\}\,, \quad \mu \neq 0\,, \qquad (1.37)$$

with $\mathcal{I}$ denoting the identity operator. Generally, time-periodic DDEs have infinitely many characteristic multipliers. If μ is a characteristic multiplier, and $\mu = \mathrm{e}^{\lambda T}$, then λ is called the characteristic exponent. Similarly to linear periodic ODEs, each characteristic multiplier is associated with infinitely many characteristic exponents of the form $\lambda_k = \gamma + \mathrm{i}(\omega + k2\pi/T)$, where $\gamma, \omega \in \mathbb{R}$, $k \in \mathbb{Z}$, and $T\omega \in (-\pi, \pi]$.

A necessary and sufficient condition for the asymptotic stability of the time-periodic DDE (1.27) with (1.28) is that all the characteristic multipliers have modulus less than one (that is, all the characteristic exponents have negative real parts) and there exist a scalar $\nu > 0$ such that

$$\int_{-\infty}^{0} \mathrm{e}^{-\nu\vartheta} \left| \mathrm{d}_\vartheta \eta_{jk}(t, \vartheta) \right| < \infty\,, \quad j, k = 1, 2, \ldots, n\,, \qquad (1.38)$$

for all $t \in \mathbb{R}$, where $\eta_{jk}(t, \vartheta)$ are the elements of $\boldsymbol{\eta}(t, \vartheta)$. Similarly to autonomous DDEs, condition (1.38) trivially holds if σ in the lower limit of the integral in (1.28) is finite. The difficulty in the stability analysis of periodic DDEs is that the monodromy operator $\mathcal{U}(T)$ has generally no closed form; consequently, the stability conditions cannot be derived in an analytic form.

There exist several numerical and semi-analytical techniques to determine the stability conditions for periodic DDEs. Most of them were developed with the aim of constructing stability charts for milling processes. Budak and Altintas [39, 40] and Merdol and Altintas [186] developed the multi-frequency solution method, which is a kind of alternative application of Hill's infinite determinant method. Butcher et al. [44, 43] used an expansion of the solution in terms of Chebyshev polynomials to obtain an approximate monodromy matrix. A temporal finite element method using Hermite polynomials was developed by Bayly et al. [23] for interrupted turning, and formulated to general periodic DDEs by Mann and Patel [179]. Szalai et al. [272] used the characteristic matrices of the system to derive stability charts (see also [244]). Recently, Khasawneh and Mann presented an effective numerical algorithm called the spectral element method that is a temporal finite element method involving highly accurate numerical quadratures for the integral terms [147, 148].

The semi-discretization method was introduced by Insperger and Stepan [123] for general time-periodic DDEs. This method was later referred to as zeroth-order semi-discretization. Elbeyli and Sun [73] generalized the method to the so-called improved zeroth-order (see also [126]) and the first-order semi-discretization. The general higher-order formalism was presented in [133]. The convergence of the method was established by Hartung et al. [101] for a large class of DDEs appearing in engineering applications. It was shown that semi-discretization preserves asymptotic stability of the original equation; therefore it can be used to construct approximate stability charts.

The merit of the semi-discretization method is that it can be used effectively to determine stability charts for time-periodic DDEs arising in different engineering problems while the numerical scheme itself is relatively simple. One of the main fields of application of semi-discretization is the stability prediction for machining processes. Different milling models were analyzed by Gradišek et al. [90], Henninger and Eberhard [105, 106], Sims et al. [249], Dombovari et al. [69], Bachrathy et al. [16], and Wan et al. [293] using the semi-discretization method. Ding et al. [65, 66] applied an alternative semi-discretization method for the milling problem using a slightly different concept in the discretization scheme (see also [119]). Sellmeier and Denkena [237] applied the method to milling processes with uneven tooth pitches and pointed out that the method can be considered an extension of the theory of sampled control systems with feedback delay according to Ackermann [1] (see also [162, 15, 209]).

A continuous-time approximation technique was introduced by Sun and Song [267, 268] based on the concept of semi-discretization that can handle multiple time delays for both linear and nonlinear dynamical systems. Models with time-periodic delays were considered by Insperger and Stepan [125], Long et al. [172], Faassen et al. [80], Zatarain et al. [304], and Seguy et al. [236]. The method can also be used for other models, for instance, for the stability analysis of periodic control systems with delayed feedback, as was shown by Sheng et al. [241] and by Konishi and Hara [154]. The application of the method is also effective for nonlinear systems at their periodic solutions where the periodic solutions are far from harmonic, as in the case of neural systems with spiking signals [223, 216].

Chapter 2
Stability Charts for Fundamental Delay-Differential Equations

Simple scalar equations play an important role in understanding the main features of DDEs and the function of stability charts. Stability charts are diagrams constructed in the plane of two (or more) parameters of the system showing the stable and unstable domains or the numbers of unstable characteristic exponents/multipliers. In this chapter, some basic scalar equations are considered for which the stability charts can be constructed in closed form by a straightforward analysis of the characteristic equation. In Sections 2.1, 2.2, and 2.3, first- and second-order autonomous scalar DDEs are analyzed by the D-subdivision method, while in Section 2.4, a second-order time-periodic scalar DDE, the delayed Mathieu equation, is analyzed by Hill's infinite determinant method.

2.1 First-Order Scalar Equations

In this section, stability properties of linear first-order scalar DDEs with point delays and with distributed delays are analyzed. The corresponding stability charts are constructed by the D-subdivision method and by analysis of the exponent-crossing directions.

2.1.1 The Hayes Equation

Consider the first-order scalar equation with a single point delay in the form

$$\dot{x}(t) = ax(t) + bx(t - \tau) . \tag{2.1}$$

This equation is often referred to as one of the simplest basic examples of a delayed system [153, 255, 188, 251]. The stability condition for the parameters a and b was first presented by Hayes [104] in 1950. In this section, the stability chart with the numbers of unstable characteristic exponents is constructed in an analytic way.

If $b = 0$, then (2.1) reduces to the scalar ODE

$$\dot{x}(t) = ax(t) \tag{2.2}$$

with the characteristic function

$$D(\lambda) = \lambda - a . \tag{2.3}$$

In this case, the only characteristic exponent is $\lambda = a$; consequently, the system is asymptotically stable if $a > 0$, and it is unstable if $a < 0$.

The stability properties for the case $b \neq 0$ are more complex due to the infinite-dimensional nature of the delayed system. The corresponding characteristic function reads

$$D(\lambda) = \lambda - a - be^{-\lambda\tau} . \tag{2.4}$$

According to the D-subdivision method, substitution of $\lambda = \gamma \pm i\omega$, $\omega \geq 0$, into the characteristic equation $D(\lambda) = 0$ and decomposition into real and imaginary parts yields

$$\text{Re} : \quad \gamma - a - be^{-\gamma\tau} \cos(\omega\tau) = 0 , \tag{2.5}$$

$$\text{Im} : \quad \omega + be^{-\gamma\tau} \sin(\omega\tau) = 0 . \tag{2.6}$$

The case $\gamma = 0$ gives the D-curves as a parametric function of ω in the form

$$\text{if } \omega = 0 : \quad b = -a , \tag{2.7}$$

$$\text{if } \omega\tau \neq k\pi , \ k \in \mathbb{N} : \quad a = \frac{\omega \cos(\omega\tau)}{\sin(\omega\tau)} , \quad b = \frac{-\omega}{\sin(\omega\tau)} , \tag{2.8}$$

with the corresponding limits for $\omega\tau = k\pi$, $k \in \mathbb{N}$. The D-curve given by (2.7) is associated with a real critical characteristic exponent crossing the imaginary axis at 0. The D-curves given by (2.8) are associated with a complex conjugate pair of characteristic exponents in the form $\lambda = \pm i\omega$, where ω is the angular frequency of the arising vibrations.

Due to the continuity of the characteristic exponents with respect to changes in the system parameters, these curves divide the parameter plane (a, b) into regions where the numbers of unstable characteristic exponents are constant. As mentioned in Section 1.3, the change of these numbers along the D-curves can be determined by analysis of the exponent-crossing direction, which is the sign of the partial derivative of γ with respect to one of the system parameters.

Taking the partial derivative of (2.5) and (2.6) with respect to b and considering that $\gamma = 0$ along the D-curves gives

$$(1 + b\tau \cos(\omega\tau))\gamma'_b + (b\tau \sin(\omega\tau))\omega'_b = \cos(\omega\tau) , \tag{2.9}$$

$$-(b\tau \sin(\omega\tau))\gamma'_b + (1 + b\tau \cos(\omega\tau))\omega'_b = -\sin(\omega\tau) , \tag{2.10}$$

where γ'_b and ω'_b are the partial derivatives of γ and ω with respect to b. The solution of (2.9)–(2.10) for γ'_b is

$$\gamma'_b = \frac{\cos(\omega\tau) + b\tau}{(1 + b\tau\cos(\omega\tau))^2 + (b\tau\sin(\omega\tau))^2}. \tag{2.11}$$

For the D-curve (2.7), equation (2.11) gives $\gamma'_b = (1 + b\tau)^{-1}$. If $b > -1/\tau$, then γ'_b is positive; consequently, the critical characteristic exponent crosses the imaginary axis from left to right through the origin as b is increased, i.e., a stable characteristic exponent becomes unstable. If $b < -1/\tau$, then the critical characteristic exponent crosses the imaginary axis in the opposite direction, i.e., an unstable exponent becomes stable for increasing b.

For the D-curves (2.8), equation (2.11) gives $\mathrm{sgn}(\gamma'_b) = \mathrm{sgn}(\cos(\omega\tau) + b\tau)$. This yields two possible cases.

1. If $\omega \in ((2k-1)\pi, 2k\pi)$, $k \in \mathbb{Z}^+$, then the corresponding D-curves (2.8) satisfy $b > 1/\tau$; consequently, γ'_b is positive in this case. This means that the complex conjugate pair of critical characteristic exponents crosses the imaginary axis from left to right, i.e., two stable exponents become unstable as b is increased.
2. If $\omega \in (2k\pi, (2k+1)\pi)$, $k \in \mathbb{N}$, then the D-curves (2.8) satisfy $b < -1/\tau$; consequently, γ'_b is negative. This means that the complex conjugate pair of critical characteristic exponents crosses the imaginary axis from right to left, i.e., two unstable exponents become stable as b is increased.

Overall, it can be concluded that more and more unstable characteristic exponents appear along the D-curves (2.8) as $|b|$ is increased. Since the half-line $b = 0$ with $a < 0$ is associated with an asymptotically stable ODE, the number of unstable exponents in the region containing this half-line is 0. The number of unstable exponents in the neighboring domains can be given by considering the exponent-crossing directions along the D-curves. The corresponding stability chart with the number of unstable characteristic exponents is presented in Figure 2.1. Stable domains (with 0 unstable exponents) are indicated by gray shading. The limits for the frequency parameter ω along the D-curves are also presented.

The number of unstable characteristic exponents can also be determined using Stepan's formula (1.26) for systems with odd dimension. According to (1.24), the real and the imaginary part of $D(i\omega)$ define the functions

$$R(\omega) = -a - b\cos(\omega\tau), \quad \omega \in [0, \infty), \tag{2.12}$$
$$S(\omega) = \omega + b\sin(\omega\tau), \quad \omega \in [0, \infty). \tag{2.13}$$

Note that the D-curves (2.7) and (2.8) are given by the equations $R(\omega) = 0$ and $S(\omega) = 0$. The stability analysis can be performed for each domain separated by the D-curves step by step. Consider, for instance, the parameters $a = 0$, $b = -5$ and $\tau = 1$ that are associated with point A within the domain of two unstable exponents (see Figure 2.1). For these parameters, $S(\omega)$ has the zeros $\sigma_1 = 2.5957$ and $\sigma_2 = 0$, and (1.26) gives $N = 2$. The analysis for the other domains can be performed in the

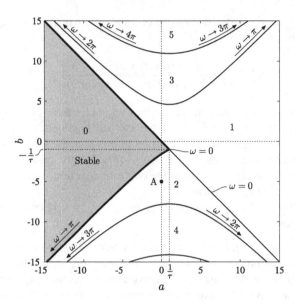

Fig. 2.1 Stability chart with the number of unstable characteristic exponents for (2.1) with $\tau = 1$.

same way, and the number of unstable exponents can be determined for the whole parameter plane (a, b).

2.1.2 The Cushing Equation

The counterpart of (2.1) with distributed delay reads

$$\dot{x}(t) = ax(t) + b \int_{-\sigma}^{0} w(\vartheta)x(t + \vartheta) \, d\vartheta . \tag{2.14}$$

This equation was analyzed by Cushing [60] in relation to population dynamics. If the kernel function is $w(\vartheta) = \delta(\vartheta - \tau)$, $\sigma \leq \tau \leq 0$, with $\delta(\vartheta)$ being the Dirac delta distribution, then (2.14) gives (2.1). Stability properties of (2.14) with the kernel function $w(\vartheta)$ being the gamma distribution was analyzed in [29, 30, 198]. Here, the stability properties are determined for the kernel function $w(\vartheta) \equiv 1$ (see [18]).

The characteristic function for (2.14) with $w(\vartheta) \equiv 1$ reads

$$D(\lambda) = \lambda - a - b \frac{1 - e^{-\lambda\sigma}}{\lambda} , \quad \lambda \neq 0 , \tag{2.15}$$

with the continuous extension at $\lambda = 0$ with

$$D(0) = \lim_{\lambda \to 0} D(\lambda) = -a - b\sigma .$$ (2.16)

According to the D-subdivision method, substitution of $\lambda = \gamma \pm i\omega$, $\omega \geq 0$, into $D(\lambda) = 0$ and decomposition into real and imaginary parts gives

$$\text{Re}: \quad \gamma - a - b\frac{\gamma - \gamma e^{-\gamma\sigma}\cos(\omega\sigma) + \omega e^{-\gamma\sigma}\sin(\omega\sigma)}{\gamma^2 + \omega^2} = 0 ,$$ (2.17)

$$\text{Im}: \quad \omega + b\frac{\omega - \omega e^{-\gamma\sigma}\cos(\omega\sigma) - \gamma e^{-\gamma\sigma}\sin(\omega\sigma)}{\gamma^2 + \omega^2} = 0 .$$ (2.18)

If $\gamma = 0$, then (2.17) and (2.18) give the D-curves in the parametric form

$$\text{if } \omega = 0: \quad b = -\frac{1}{\sigma}a ,$$ (2.19)

$$\text{if } \omega\tau \neq k\pi, \; k \in \mathbb{N}: \quad a = \frac{\omega\sin(\omega\sigma)}{1 - \cos(\omega\sigma)} , \quad b = \frac{-\omega^2}{1 - \cos(\omega\sigma)} ,$$ (2.20)

with the corresponding limits for $\omega\sigma = k\pi$, $k \in \mathbb{N}$.

Similarly to the analysis of (2.1), the exponent-crossing direction at the D-curves can be obtained by taking the partial derivatives of (2.17) and (2.18) with respect to b and then setting $\gamma = 0$ and solving the resulting equations for γ'_b. This derivation gives

$$\gamma'_b = \frac{1}{A^2 + B^2}\left(\frac{1 - \cos(\omega\sigma)}{\omega^2}b\sigma + \frac{\sin(\omega\sigma)}{\omega}\right) ,$$ (2.21)

where

$$A = 1 - \frac{1 - \cos(\omega\sigma) - \omega\sigma\sin(\omega\sigma)}{\omega^2}b ,$$ (2.22)

$$B = \frac{\sin(\omega\sigma) - \omega\sigma\cos(\omega\sigma)}{\omega^2}b .$$ (2.23)

Taking the limit $\omega \to 0$ gives the exponent-crossing direction along the D-curve (2.19) associated with $\omega = 0$ as

$$\gamma'_b = \frac{2\sigma}{b\sigma^2 + 2} .$$ (2.24)

If $b > -2/\sigma^2$, then γ'_b is positive; consequently, as b is increased, the critical characteristic exponent crosses the imaginary axis from left to right through the origin, i.e., a stable exponent becomes unstable. If $b < -2/\sigma^2$, then γ'_b is negative, and the critical characteristic exponent crosses the imaginary axis in the opposite direction, i.e., an unstable exponent becomes stable as b is increased.

For the D-curves (2.20), equation (2.21) gives

$$\text{sgn}\,\gamma'_b = \text{sgn}\left(-\sigma + \frac{\sin(\omega\sigma)}{\omega}\right) ,$$ (2.25)

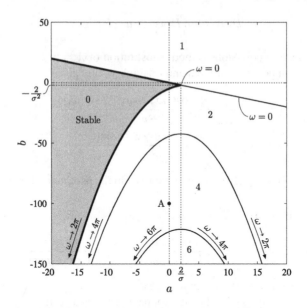

Fig. 2.2 Stability chart with the number of unstable characteristic exponents for (2.14) with $w(\vartheta) \equiv 1, \sigma = 1$.

which is always negative if $\omega \neq 0$. This implies that new pairs of unstable exponents appear at each D-curve given by (2.20) as b is decreased.

Utilizing that the half-line $b = 0$ with $a < 0$ is associated with an asymptotically stable ODE, the number of unstable exponents can be given for each region by considering the exponent-crossing direction at the corresponding D-curves. The stability chart with the number of unstable exponents can be seen in Figure 2.2. Stable domains are indicated by gray shading. The limits for the frequency parameter ω along the D-curves are also presented.

Again, the application of Stepan's formula (1.26) with the functions

$$R(\omega) = -a - b\frac{\sin(\omega\sigma)}{\omega}, \quad \omega \in [0, \infty), \tag{2.26}$$

$$S(\omega) = \omega + b\frac{1 - \cos(\omega\sigma)}{\omega}, \quad \omega \in [0, \infty), \tag{2.27}$$

defined in (1.24) can also be used to determine the number of unstable characteristic exponents for the individual domains in Figure 2.2. Consider, for instance, the parameters $a = 0$, $b = -100$, and $\sigma = 1$ that are associated with point A within the domain of four unstable exponents. For these parameters, $S(\omega)$ has the zeros $\sigma_1 = 10.823$, $\sigma_2 = 7.3814$, $\sigma_3 = 5.4864$, $\sigma_4 = 0$, and (1.26) gives $N = 4$. The analysis for the other domains can be performed in the same way.

2.2 Delayed Oscillators

A well-known Newtonian example for delayed systems is the damped delayed oscillator described by the scalar DDE

$$\ddot{x}(t) + a_1 \dot{x}(t) + a_0 x(t) = b_0\, x(t - \tau)\,. \tag{2.28}$$

Although the stability chart in the parameter plane (a_0, b_0) has a very clear structure (see Figures 2.4 and 2.5), it was first published correctly only in 1966 by Hsu and Bhatt [112]. Since then, this equation has become a basic example for delayed Newtonian problems; see, for instance, [153, 255, 44, 179]. Stability analysis for the generalizations of (2.28) have been considered in detail in [53, 160, 245, 176].

The special case of (2.28) with $b_0 = 0$ is the damped oscillator

$$\ddot{x}(t) + a_1 \dot{x}(t) + a_0 x(t) = 0\,. \tag{2.29}$$

The characteristic function (which is a polynomial in this case) reads

$$D(\lambda) = \lambda^2 + a_1 \lambda + a_0\,, \tag{2.30}$$

and the characteristic exponents are

$$\lambda_{1,2} = \frac{-a_1 \pm \sqrt{a_1^2 - 4a_0}}{2}\,. \tag{2.31}$$

Substitution of $\lambda = \pm i\omega$, $\omega \geq 0$, into (2.30) gives the D-curves in the form

$$\text{if } \omega = 0 :\quad a_0 = 0\,,\quad a_1 \in \mathbb{R}\,, \tag{2.32}$$

$$\text{if } \omega \neq 0 :\quad a_1 = 0\,,\quad a_0 = \omega^2 > 0\,. \tag{2.33}$$

The number of unstable exponents can be determined based on (2.31). The stability properties are presented in Figure 2.3.

Consider now the undamped delayed oscillator

$$\ddot{x}(t) + a_0 x(t) = b_0 x(t - \tau)\,, \tag{2.34}$$

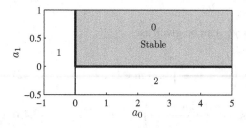

Fig. 2.3 Stability chart with the number of unstable characteristic exponents for the damped oscillator (2.29).

which is a special case of (2.28) with $a_1 = 0$. The corresponding characteristic function is

$$D(\lambda) = \lambda^2 + a_0 - b_0 e^{-\lambda \tau} \, . \tag{2.35}$$

According to the D-subdivision method, substitution of $\lambda = \gamma \pm i\omega$, $\omega \geq 0$, into $D(\lambda) = 0$ and decomposition into real and imaginary parts yields

$$\text{Re} \; : \quad \gamma^2 - \omega^2 + a_0 - b_0 e^{-\gamma \tau} \cos(\omega \tau) = 0 \, , \tag{2.36}$$

$$\text{Im} \; : \quad 2\gamma\omega + b_0 e^{-\gamma \tau} \sin(\omega \tau) = 0 \, . \tag{2.37}$$

If $\gamma = 0$, then (2.36) and (2.37) give the D-curves in the form

$$\text{if } \omega \tau = k\pi \, , \; k \in \mathbb{N} \; : \quad b_0 = (-1)^k a_0 - (-1)^k \left(\frac{k\pi}{\tau} \right)^2 \, , \tag{2.38}$$

$$\text{if } \omega \tau \neq k\pi \, , \; k \in \mathbb{N} : \quad b_0 = 0 \, , \quad a_0 = \omega^2 > 0 \, . \tag{2.39}$$

These D-curves are straight lines in the plane (a_0, b_0) and form a special combination of triangles, shown in Figure 2.4. The D-curve $b_0 = a_0$ given by (2.38) with $k = 0$ is associated with a real critical characteristic exponent $\lambda = 0$. All the other D-curves are associated with the complex conjugate pair of characteristic exponents of the form $\lambda = \pm i\omega$.

The exponent-crossing direction along the D-curves can be obtained by taking the partial derivatives of (2.36) and (2.37) with respect to b_0 and substituting the D-curves (2.38), (2.39), and $\gamma = 0$. For the D-curves (2.38), this analysis gives

$$\gamma'_{b_0} = \frac{b_0 \tau}{b_0^2 \tau^2 + 4\omega^2} \, , \tag{2.40}$$

that is, γ'_{b_0} is positive for $b_0 > 0$ and negative for $b_0 < 0$. This means that new unstable exponents appear as these D-curves are crossed with increasing $|b_0|$. Along the D-curve $a_0 = b_0$ associated with $k = 0$ in (2.38), a real characteristic exponent becomes unstable as $|b_0|$ is increased, while at the other D-curves associated with $k > 0$, a pair of complex characteristic exponents crosses the imaginary axis from left to right for increasing $|b_0|$.

The analysis for the D-curve (2.39) gives

$$\gamma'_{b_0} = \frac{-\sin(\omega \tau)}{2\omega} \tag{2.41}$$

with $\omega = \sqrt{a_0}$. In this case, the sign of γ'_{b_0} can be given as

$$\gamma'_{b_0} < 0 \quad \text{if} \quad \left(\frac{2m\pi}{\tau} \right)^2 < a_0 < \left(\frac{(2m+1)\pi}{\tau} \right)^2 \, , \quad m \in \mathbb{N} \, , \tag{2.42}$$

$$\gamma'_{b_0} > 0 \quad \text{if} \quad \left(\frac{(2m+1)\pi}{\tau} \right)^2 < a_0 < \left(\frac{(2m+2)\pi}{\tau} \right)^2 \, , \quad m \in \mathbb{N} \, . \tag{2.43}$$

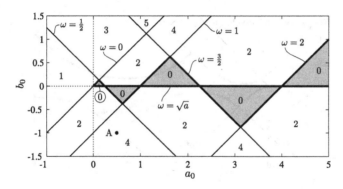

Fig. 2.4 Hsu–Bhatt stability chart with the number of unstable characteristic exponents for (2.34) with $\tau = 2\pi$.

This means that a pair of complex characteristic exponents becomes unstable as the line $b_0 = 0$ with $a_0 > 0$ is crossed from inside the triangles (see Figure 2.4).

Since the half-line $b_0 = 0$ with $a_0 < 0$ is associated with an ODE with one unstable characteristic exponent (see Figure 2.3), the number of unstable exponents can be given for all regions by considering the exponent-crossing directions along the corresponding D-curves. The stability chart with the number of unstable characteristic exponents is presented in Figure 2.4. Stable domains are indicated by gray shading. The frequency parameter ω along the stability boundaries is also presented.

The number of unstable characteristic exponents can also be determined using Stepan's formula (1.25) for systems with even dimension. According to (1.24), the real and the imaginary part of $D(i\omega)$ define the functions

$$R(\omega) = -\omega^2 + a_0 - b_0 \cos(\omega\tau) , \quad \omega \in [0, \infty) , \tag{2.44}$$
$$S(\omega) = b_0 \sin(\omega\tau) , \quad \omega \in [0, \infty) . \tag{2.45}$$

For this analysis, (1.25) should be applied for some fixed parameters associated with the individual domains in Figure 2.4. For instance, the parameters $a_0 = 0.5$, $b_0 = -1$, and $\tau = 2\pi$ are associated with point A within the domain of four unstable exponents. For these parameters, $R(\omega)$ has the zeros $\rho_1 = 1.1161, \rho_2 = 0.7631$, $\rho_3 = 0.3156$, and (1.25) gives $N = 4$. All the other domains can be analyzed in a similar way.

Consider now the damped and delayed case, i.e., (2.28) with $a_1 \neq 0$ and $b_0 \neq 0$. The corresponding characteristic function is

$$D(\lambda) = \lambda^2 + a_1\lambda + a_0 - b_0 e^{-\lambda\tau} . \tag{2.46}$$

Substitution of $\lambda = \pm i\omega, \omega \geq 0$, into $D(\lambda) = 0$ gives the D-curves in the form

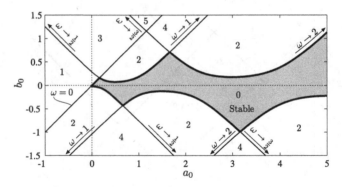

Fig. 2.5 Hsu–Bhatt stability chart with the number of unstable characteristic exponents for (2.28) with $a_1 = 0.1$ and $\tau = 2\pi$.

$$\text{if } \omega = 0 \ : \quad b_0 = a_0 \ , \tag{2.47}$$

$$\text{if } \omega\tau \neq k\pi \ , \ k \in \mathbb{N} \ : \quad a_0 = \omega^2 + \frac{a_1\omega\cos(\omega\tau)}{\sin(\omega\tau)} \ , \quad b_0 = \frac{-a_1\omega}{\sin(\omega\tau)} \ . \tag{2.48}$$

The number of unstable exponents for the domains bounded by (2.47) and (2.48) can be determined based on continuous dependence of the characteristic exponents on the system parameters. For this purpose, the stability chart for the case $a_1 = 0$ in Figure 2.4 can be used as a reference. The corresponding stability chart is shown in Figure 2.5. Stable domains are indicated by gray shading. The limits for the frequency parameter ω along the stability boundaries are also presented.

2.3 Stabilization with Feedback Delay

Stabilization of a one-degree-of-freedom Newtonian system about an unstable equilibrium by a proportional-derivative controller in the presence of feedback delay is described by the equation

$$\ddot{x}(t) + a_0 x(t) = -k_p x(t - \tau) - k_d \dot{x}(t - \tau) \ , \tag{2.49}$$

where $a_0 < 0$ is the negative stiffness, k_p is the proportional gain, k_d is the derivative gain, and τ is the feedback delay. This equation corresponds to the general second-order delayed system

$$\ddot{x}(t) + a_1 \dot{x}(t) + a_0 x(t) = b_0 x(t - \tau) + b_1 \dot{x}(t - \tau) \tag{2.50}$$

with $b_0 = -k_p$, $b_1 = -k_d$, and $a_1 = 0$. Equation (2.49) is a frequently cited example in dynamics and control theory [255, 243, 166], and it is also relevant to human balancing problems [263, 260, 190].

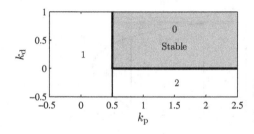

Fig. 2.6 Stability chart with the number of unstable characteristic exponents for (2.51) with $a_0 = -0.5$.

If $a_0 < 0$ and $k_p = 0$, $k_d = 0$, then the system is unstable with the characteristic exponents $\lambda_{1,2} = \pm \sqrt{-a_0}$, i.e., the number of unstable exponents is 1.

If the feedback delay is zero, then the governing equation reads

$$\ddot{x}(t) + a_0 x(t) = -k_p x(t) - k_d \dot{x}(t) , \qquad (2.51)$$

which is equivalent to the well-known damped oscillator (2.29) with some appropriate parameter transformations. In this case, the D-curves can be given in the form

$$\text{if } \omega = 0 : \quad k_p = a_0 , \quad k_d \in \mathbb{R} , \qquad (2.52)$$

$$\text{if } \omega \neq 0 : \quad k_d = 0 , \quad k_p = -a_0 + \omega^2 > -a_0 , \qquad (2.53)$$

where ω is the imaginary part of the critical characteristic exponent. The stability chart with the number of unstable exponents is presented in Figure 2.6. This chart shows that if the feedback delay τ is 0, then the unstable system can be stabilized by control gains $k_p > -a_0$ and $k_d > 0$.

Consider now (2.49) with $\tau > 0$. The corresponding characteristic function reads

$$D(\lambda) = \lambda^2 + a_0 + k_p e^{-\lambda\tau} + k_d \lambda e^{-\lambda\tau} . \qquad (2.54)$$

According to the D-subdivision method, substitution of $\lambda = \gamma \pm i\omega$, $\omega \geq 0$, into $D(\lambda) = 0$ and decomposition into real and imaginary parts gives

$$\text{Re} : \quad \gamma^2 - \omega^2 + a_0 + k_p e^{-\gamma\tau} \cos(\omega\tau) + k_d \gamma e^{-\gamma\tau} \cos(\omega\tau) + k_d \omega e^{-\gamma\tau} \sin(\omega\tau) = 0 , \qquad (2.55)$$

$$\text{Im} : \quad 2\gamma\omega - k_p e^{-\gamma\tau} \sin(\omega\tau) + k_d \omega e^{-\gamma\tau} \cos(\omega\tau) - k_d \gamma e^{-\gamma\tau} \sin(\omega\tau) = 0 . \qquad (2.56)$$

The case $\gamma = 0$ gives the D-curves in the form

$$\text{if } \omega = 0 : \quad k_p = -a_0 , \quad k_d \in \mathbb{R} , \qquad (2.57)$$

$$\text{if } \omega \neq 0 : \quad k_p = (\omega^2 - a_0) \cos(\omega\tau) , \quad k_d = \frac{\omega^2 - a_0}{\omega} \sin(\omega\tau) . \qquad (2.58)$$

The D-curve $k_p = -a_0$ given by (2.57) is associated with a real critical characteristic exponent $\lambda = 0$. The D-curve given by (2.58) is associated with a complex conjugate

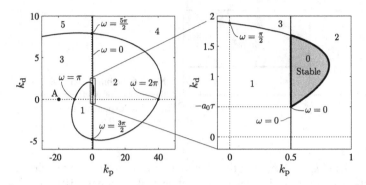

Fig. 2.7 Stability chart with the number of unstable characteristic exponents for (2.49) with $a_0 = -0.5$ and $\tau = 1$.

pair of characteristic exponents of the form $\lambda = \pm i\omega$. For a fixed a_0, these curves cut the parameter plane (k_p, k_d) into infinitely many domains (see Figure 2.7).

The exponent-crossing direction along the D-curve $k_p = -a_0$ can be obtained by taking the partial derivatives of (2.55) and (2.56) with respect to k_p and substituting $\gamma = 0$, $\omega = 0$, and $k_p = -a_0$. This derivation gives

$$\gamma'_{k_p} = -\frac{1}{a_0\tau + k_d}, \tag{2.59}$$

that is, γ'_{k_p} is positive for $k_d < -a_0\tau$ and negative for $k_d > -a_0\tau$. If the line $k_p = -a_0$ is crossed from left to right with $k_d > -a_0\tau$, then a real characteristic exponent becomes stable. If $k_d < -a_0\tau$, then a real exponent becomes unstable as the line $k_p = -a_0$ is crossed from left to right. Since the parameter point $(k_p, k_d) = (0, 0)$ corresponds to an ODE with one unstable characteristic exponent, the number of unstable exponents can be given for all the domains in the parameter plane (k_p, k_d) by considering the exponent-crossing directions along the D-curve $k_p = -a_0$. The corresponding stability chart with the number of unstable exponents is presented in Figure 2.7. The stable domain is indicated by gray shading. Some specific values of the frequency parameter ω along the stability boundaries are also presented.

The number of unstable exponents can also be determined using Stepan's formula (1.25) with the functions

$$R(\omega) = -\omega^2 + a_0 + k_p \cos(\omega\tau) + k_d\omega \sin(\omega\tau), \quad \omega \in [0, \infty), \tag{2.60}$$

$$S(\omega) = -k_p \sin(\omega\tau) - k_d\omega \cos(\omega\tau), \quad \omega \in [0, \infty). \tag{2.61}$$

For instance, the parameters $a_0 = -0.5$, $\tau = 1$, $k_p = -20$, and $k_d = 0$ are associated with point A within the domain of three unstable exponents in Figure 2.7. For these parameters, $R(\omega)$ has the zeros $\rho_1 = 3.844$ and $\rho_2 = 1.750$, and (1.25) gives $N = 3$. All the other domains can be analyzed in a similar way.

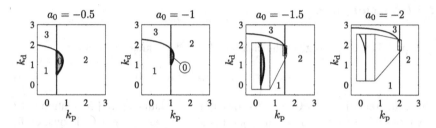

Fig. 2.8 Stability charts with the number of unstable characteristic exponents for (2.49) with $\tau = 1$ for different system parameters a_0.

Figure 2.8 shows the stability charts for different system parameters a_0. It can be observed that as a_0 is decreased, the stable domain shrinks, and at $a_0 = -2$, it completely disappears. This can easily be seen by the analysis of the tangent of the parametric curve (2.58) at $\omega = 0$. A long but straightforward analysis gives

$$\lim_{\omega \to 0} \frac{dk_d}{dk_p} = \lim_{\omega \to 0} \frac{\dfrac{dk_d}{d\omega}}{\dfrac{dk_p}{d\omega}} = \frac{a_0\tau^3 + 6\tau}{3a_0\tau^2 + 6}. \tag{2.62}$$

The tangent is vertical if $3a_0\tau^2 + 6 = 0$, which gives the critical value $a_{0,\text{crit}} = -2/\tau^2$. If $a_0 < a_{0,\text{crit}}$, then (2.49) is unstable for any k_p and k_d. For the case $\tau = 1$ in Figure 2.8, the critical value is $a_{0,\text{crit}} = -2$.

The same phenomenon is often considered from the delay point of view: for a fixed system parameter $a_0 < 0$, there exists a critical delay $\tau_{\text{crit}} = \sqrt{-2/a_0}$. If the feedback delay is larger than τ_{crit}, then the system is unstable for all combinations of k_p and k_d.

2.4 Delayed Mathieu Equation

A paradigm for time-delayed and time-periodic systems is the damped and delayed Mathieu equation

$$\ddot{x}(t) + a_1\dot{x}(t) + (\delta + \varepsilon \cos t)\, x(t) = b_0 x(t - \tau). \tag{2.63}$$

This equation combines the phenomenon of parametric forcing with the effect of time delay. Consequently, stability analysis can be performed based on the Floquet theory of DDEs [97, 100]. Here, the special resonant case will be investigated, when $\tau = 2\pi$, i.e., the time delay is just equal to the principal period, and the stability charts will be constructed by Hill's infinite determinant method.

2.4.1 Special Cases

The special case $b_0 = 0$ gives the classical damped Mathieu equation

$$\ddot{x}(t) + a_1 \dot{x}(t) + (\delta + \varepsilon \cos t)\, x(t) = 0 \,. \tag{2.64}$$

This equation (with $a_1 = 0$) was first discussed by Mathieu [183] in relation to the vibrations of an elliptic membrane, but it is also known as the model equation for the small oscillations of a pendulum under parametric forcing (i.e., with periodically vibrating suspension point) around its upper and the lower equilibria [265, 146, 170]. The stability chart in the plane (δ, ε), the so-called Ince–Strutt diagram, was published by van der Pol and Strutt [286] in 1928, but the functions of the stability boundaries were already given in the textbook of Ince [116] in 1926.

According to the Floquet theory, the stability of (2.64) is determined by the eigenvalues of the monodromy matrix. As was mentioned in Section 1.2, this matrix has, in general, no closed-form representation, but there exist several methods to give a sufficiently accurate approximation (see, for instance, [205] for an overview). Here the piecewise constant approximation of the periodic matrix is used according to [111]. Equation (2.64) can be transformed into the state-space form

$$\dot{\mathbf{y}}(t) = \mathbf{A}(t)\mathbf{y}(t), \quad \mathbf{A}(t) = \mathbf{A}(t + T) \,, \tag{2.65}$$

with

$$\mathbf{y}(t) = \begin{pmatrix} x(t) \\ \dot{x}(t) \end{pmatrix}, \quad \mathbf{A}(t) = \begin{pmatrix} 0 & 1 \\ -(\delta + \varepsilon \cos t) & -a_1 \end{pmatrix} \,. \tag{2.66}$$

After piecewise constant approximation of the coefficient matrix according to (1.9), the Floquet transition matrix can be approximated as

$$\mathbf{\Phi}(T) \approx \tilde{\mathbf{\Phi}}(T) = e^{\mathbf{A}_{p-1}h}\, e^{\mathbf{A}_{p-2}h} \cdots e^{\mathbf{A}_0 h} \,, \tag{2.67}$$

where $h = T/p$ is the length of the discretization step, with p being an integer, and $\mathbf{A}_i$ is the average of the time-periodic matrix $\mathbf{A}(t)$ over the ith discretization step defined by (1.8). The system is asymptotically stable if the eigenvalues of $\mathbf{\Phi}(T)$ (i.e., the characteristic multipliers) have modulus less than one. The stability chart can be constructed by point-by-point evaluation of the critical eigenvalues over a fixed-sized grid of parameters δ and ε. The corresponding stability chart obtained by $p = 20$ is shown in Figure 2.9 for different damping parameters a_1.

According to Liouville's formula [83], the determinant of the monodromy matrix can be given as

$$\det \mathbf{\Phi}(T) = \exp\left(\int_0^T \mathrm{Tr}\, \mathbf{A}(t)\, dt \right) = \exp\left(\int_0^T -a_1\, dt \right) = e^{-a_1 T} \,. \tag{2.68}$$

Since $\mathbf{A}$ is a 2×2 matrix, $\det \mathbf{\Phi}(T) = \mu_1 \mu_2$, where μ_1 and μ_2 are the two characteristic multipliers with $|\mu_1| \geq |\mu_2|$. The location of the multipliers in the complex plane can be characterized as follows.

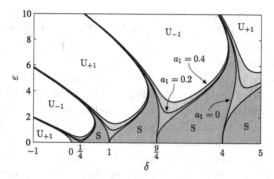

Fig. 2.9 Ince–Strutt [116, 286] stability chart for (2.64). S indicates stable domains, U_{+1} and U_{-1} indiciate unstable domains with critical characteristic multipliers $\mu_1 > 1$ and $\mu_1 < -1$, respectively. Stability boundaries are presented for $a_1 = 0$, 0.2, and 0.4.

1. If $a_1 < 0$, then $\mu_1\mu_2 > 1$, i.e., at least one of the characteristic multipliers has modulus larger than 1. Consequently, the Mathieu equation (2.64) with negative damping a_1 is always unstable.
2. If $a_1 = 0$, then $\mu_1\mu_2 = 1$. In this case, the system is stable in the Lyapunov sense if $|\mu_1| = 1$ and $|\mu_2| = 1$; otherwise, the system is unstable with $|\mu_1| > 1$ and $|\mu_2| < 1$. Asymptotic stability does not occur in this case.
3. If $a_1 > 0$, then $\mu_1\mu_2 < 1$. In this case, one of the characteristic multipliers is always located within the unit circle. The system can be asymptotically stable with $|\mu_2| \leq |\mu_1| < 1$.

The above cases show that both characteristic multipliers can never cross the unit circle at the same time. Consequently, in the damped Mathieu equation (2.64), only cyclic-fold or period-doubling (flip) bifurcations may occur, and secondary Hopf bifurcation does not arise. In Figure 2.9, stability curves bounding the unstable domains indicated by U_{+1} represent cyclic-fold bifurcations, while the curves bounding the domains indicated by U_{-1} represent period-doubling bifurcations.

The special case $\varepsilon = 0$ of (2.63) gives the damped and delayed oscillator (2.28) with $a_0 = \delta$ discussed in Section 2.2. The corresponding stability charts can be seen in Figure 2.4 for $a_1 = 0$, $\tau = 2\pi$ and in Figure 2.5 for $a_1 = 0.1$, $\tau = 2\pi$. These charts were published first by Hsu and Bhatt [112] in 1966.

2.4.2 D-curves

First, the D-curves for the undamped system in the plane (δ, b_0) are determined by Hill's infinite determinant method. The undamped case of (2.63) with $a_1 = 0$ reads

$$\ddot{x}(t) + (\delta + \varepsilon \cos t) x(t) = b_0 x(t - 2\pi) . \tag{2.69}$$

The corresponding stability chart in the space of system parameters δ, b_0, and ε was published by Insperger and Stepan [124] in 2002. This chart combines the Ince–Strutt stability chart of the Mathieu equation (shown in Figure 2.9) and the Hsu–Bhatt stability chart of the delayed oscillator (shown in Figure 2.4).

The main point in the construction of the stability boundaries for (2.69) is that the points where the lines of slope ± 1 intersect the line $b_0 = 0$ in the Hsu–Bhatt stability chart in Figure 2.4 are $a_0 = 0, 1/4, 1, 9/4, \ldots$, which coincide with the points where the unstable tongues in the Ince–Strutt stability chart in Figure 2.9 open on the δ-axis for $\varepsilon = 0$. In the subsequent analysis, this property will be used to show how the stable triangles in the plane (a_0, b_0) in Figure 2.4 change for $\varepsilon > 0$. The investigation is based on the following theorems of the Floquet theory of DDEs (see, e.g., [83]):

- The trivial solution of (2.69) is asymptotically stable if and only if all the (infinitely many) characteristic multipliers have modulus less than one;
- $\mu = e^{\lambda T}$ is a characteristic multiplier of (2.69) if and only if there exists a nontrivial solution in the form $x(t) = p(t)e^{\lambda t}$, where $p(t) = p(t + T)$, with T being the principal period (in this case $T = 2\pi$).

This implies the application of the trial solution in the form

$$x(t) = p(t)e^{\lambda t} + \overline{p}(t)e^{\overline{\lambda} t}, \tag{2.70}$$

where $p(t) = p(t + 2\pi)$ is a periodic function and bar denotes complex conjugate. Note that λ is a characteristic exponent, that is, if $\mathrm{Re}\,\lambda < 0$, then the trivial solution $x(t) \equiv 0$ is asymptotically stable.

According to Hahn [96], equation (2.69) may have solutions of the form $t^k p(t)e^{\lambda t}$, $k \in \mathbb{Z}^+$, in critical cases. Consequently, if $|\mu| = 1$, i.e., $\mathrm{Re}\,\lambda = 0$, then the solution $p(t)e^{0t}$ is stable in the Lyapunov sense, but the solutions $t^k p(t)e^{0t}$ are unstable. This case has no importance here, since it may arise only at certain special points of the stability boundaries, while in the present investigation, the domains of asymptotic stability are determined.

Expand the periodic function $p(t)$ in (2.70) into a Fourier series to obtain

$$x(t) = \left(\sum_{k=0}^{\infty} A_k e^{ikt} + B_k e^{-ikt}\right) e^{\lambda t} + \left(\sum_{k=0}^{\infty} \overline{A}_k e^{-ikt} + \overline{B}_k e^{ikt}\right) e^{\overline{\lambda} t}. \tag{2.71}$$

Using trigonometric transformations, (2.71) can be transformed to

$$x(t) = \sum_{k=-\infty}^{\infty} C_k e^{(\lambda+ik)t} + \overline{C}_k e^{(\overline{\lambda}-ik)t}. \tag{2.72}$$

Substitution into (2.69) and balancing of the harmonics $e^{(\lambda+ik)t}$ and $e^{(\overline{\lambda}-ik)t}$ yields two systems of equations for the coefficients C_k and $\overline{C}_k$:

$$\frac{\varepsilon}{2}C_{k-1} + c_k C_k + \frac{\varepsilon}{2}C_{k+1} = 0, \quad k \in \mathbb{Z}, \tag{2.73a}$$

$$\frac{\varepsilon}{2}\overline{C}_{k-1} + \overline{c}_k \overline{C}_k + \frac{\varepsilon}{2}\overline{C}_{k+1} = 0, \quad k \in \mathbb{Z}, \tag{2.73b}$$

where

$$c_k = \delta + (\lambda + ik)^2 - b_0 e^{-2\pi(\lambda+ik)}. \tag{2.74}$$

As follows from the Floquet theory, (2.73a) and (2.73b) are satisfied if and only if λ is a characteristic exponent. Since (2.73a) and (2.73b) are equivalent, only (2.73a) will be analyzed. There is a nontrivial solution of system (2.73a) if the number zero is an eigenvalue of the so-called Hill's infinite matrix

$$\mathcal{H}(\lambda) = \begin{pmatrix} \ddots & \ddots & \ddots & \ddots & & \\ \ddots & \varepsilon/2 & c_{-1} & \varepsilon/2 & 0 & \\ & 0 & \varepsilon/2 & c_0 & \varepsilon/2 & 0 \\ & & 0 & \varepsilon/2 & c_1 & \varepsilon/2 & \ddots \\ & & & \ddots & \ddots & \ddots & \ddots \end{pmatrix}. \tag{2.75}$$

This matrix represents an unbounded linear operator $\mathcal{H} : l_2^{\mathbb{Z}} \to l_2^{\mathbb{Z}}$. Here, $l_2^{\mathbb{Z}}$ is the Hilbert space of the complex sequences $(\ldots, z_{-1}, z_0, z_1, \ldots)$ with $\sum_{k=-\infty}^{\infty} |z_k|^2 < \infty$. As is the case for (unbounded) linear operators with compact resolvents, the spectrum of $\mathcal{H}$ consist of a countable number of eigenvalues. All of these eigenvalues are of finite multiplicity. The number zero is an eigenvalue of $\mathcal{H}$ if and only if

$$\text{Ker}\,\mathcal{H}(\lambda) \neq \{0\}. \tag{2.76}$$

Formula (2.76) can be treated as the characteristic equation of (2.69), since its roots are the characteristic exponents. This is a reformulation of (1.37) with $\mu = \exp(2\pi\lambda)$.

In order to carry out calculations, only the truncated system of equations with $k = -N, \ldots, N$ is considered. This reduces the infinite eigenvalue problem of operator $\mathcal{H}$ to the calculation of a finite determinant

$$D(\lambda) = \det \begin{pmatrix} c_{-N} & \varepsilon/2 & & & \\ \varepsilon/2 & c_{-N+1} & \varepsilon/2 & & \\ & \ddots & \ddots & \ddots & \\ & & \varepsilon/2 & c_{N-1} & \varepsilon/2 \\ & & & \varepsilon/2 & c_N \end{pmatrix}. \tag{2.77}$$

Although this truncation seems to be a rough approximation, it still has a sound mathematical basis (see [184, 63]). This approximation is just the same as the one applied during the construction of the Ince–Strutt diagram using Hill's method. The equation $D(\lambda) = 0$ can therefore be considered an approximate characteristic equation. In what follows, first, the D-curves will be constructed for $N \to \infty$; then, in Section 2.4.3, the domains of stability will be determined.

According to the D-subdivision method, the substitution of $\lambda = \pm i\omega$, $\omega \geq 0$, into $D(\lambda) = 0$ gives an implicit form for the approximate D-curves (or D-surfaces) of (2.69) in the parameter space $(\delta, b_0, \varepsilon)$ with the frequency parameter ω. In this case, the diagonal elements in (2.77) read

$$c_k = \delta - (\omega + k)^2 - b_0 e^{-i2\pi\omega} , \qquad k = -N, \dots, N . \qquad (2.78)$$

Note that the imaginary part of c_k does not depend on k:

$$\operatorname{Im} c_k = b_0 \sin(2\pi\omega) , \qquad k = -N, \dots, N . \qquad (2.79)$$

It can be seen that $\operatorname{Im} c_k = 0$ if and only if either $\omega = j/2$ with $j = 0, 1, \dots,$ or $b_0 = 0$, which gives the classical Mathieu equation. Now the cases $\omega \neq j/2$ and $\omega = j/2$ with $j = 0, 1, \dots$ are investigated separately.

Case $\omega \neq j/2$, $j = 0, 1, \dots$

In this case, $\operatorname{Im} c_k \neq 0$ for any k (if $b_0 \neq 0$), as follows from (2.79). The Gaussian elimination algorithm can be applied for the tridiagonal matrix in (2.77) to transform it into an upper triangular matrix having the elements d_k on the main diagonal. Clearly, $d_{-N} = c_{-N} \neq 0$. In the $(N + k)$th step of the Gaussian elimination process, Hill's matrix assumes the form

$$\begin{pmatrix} d_{-N} & \varepsilon/2 & 0 & \cdots & & & & \\ & \ddots & \ddots & \ddots & & & & \\ \cdots & 0 & d_{k-1} & \varepsilon/2 & 0 & \cdots & & \\ \cdots & 0 & d_k & \varepsilon/2 & 0 & \cdots & & \\ \cdots & 0 & \varepsilon/2 & c_{k+1} & \varepsilon/2 & 0 & \cdots & \\ \cdots & 0 & \varepsilon/2 & c_{k+2} & \varepsilon/2 & 0 & \cdots & \\ & & \ddots & \ddots & \ddots & \ddots & \ddots & \end{pmatrix} . \qquad (2.80)$$

Let us suppose that $\operatorname{sgn}(\operatorname{Im} d_k) = \operatorname{sgn}(\operatorname{Im} c_k)$ for some k. Since $\operatorname{Im} c_k \neq 0$, this means that $\operatorname{Im} d_k \neq 0$, i.e., $|d_k| \neq 0$. Thus, the subsequent elimination of $\varepsilon/2$ to the left from c_{k+1} leads to

$$d_{k+1} = c_{k+1} - \frac{\varepsilon^2}{4d_k} = \left(\operatorname{Re} c_{k+1} - \frac{\varepsilon^2 \operatorname{Re} d_k}{4|d_k|^2} \right) + i \left(\operatorname{Im} c_{k+1} + \frac{\varepsilon^2 \operatorname{Im} d_k}{4|d_k|^2} \right) . \qquad (2.81)$$

Consequently,

$$\operatorname{sgn}(\operatorname{Im} d_{k+1}) = \operatorname{sgn}\left(\operatorname{Im} c_{k+1} + \left(\frac{\varepsilon}{2|d_k|} \right)^2 \operatorname{Im} d_k \right) = \operatorname{sgn}(\operatorname{Im} d_k) = \operatorname{sgn}(\operatorname{Im} c_k) \neq 0 . \qquad (2.82)$$

Since $\operatorname{Im} d_{-N} = \operatorname{Im} c_{-N} = b \sin(2\pi\omega) \neq 0$, we have $\operatorname{Im} d_k \neq 0$, that is, $|d_k| \neq 0$ is true by induction.

The determinant of Hill's matrix can be calculated as the product of the diagonal elements of the upper triangular matrix. Hence,

$$D(i\omega) = \prod_{k=-N}^{N} d_k \neq 0 , \tag{2.83}$$

that is, the determinant (2.77) is never zero if $\lambda = i\omega$ with $\omega \neq j/2, j = 0, 1, \ldots$. This means that there is no nontrivial solution of system (2.73a), and there are no stability boundary curves in this case.

Case $\omega = j/2, \quad j = 0, 1, \ldots$

In this case, the diagonal elements in (2.77) are real:

$$c_k = \delta - \left(k + \tfrac{1}{2}j\right)^2 - b_0(-1)^j . \tag{2.84}$$

If j is even, that is, $j = 2h, h = 0, 1, \ldots$, then $\lambda = ih$, and the corresponding characteristic multiplier is

$$\mu = e^{ih2\pi} = e^{i2\pi} = 1 . \tag{2.85}$$

In this case, $c_k = \delta - b_0 - (k+h)^2$, and $D(\lambda) = 0$ gives the relation $f_{+1}(\delta - b_0, \varepsilon) = 0$ for the D-curves. For the case $b_0 = 0$, the relation $f_{+1}(\delta, \varepsilon) = 0$ serves as the U_{+1} D-curves of the classical Mathieu equation defined in the form $\delta = g_{+1}(\varepsilon)$. This means that the D-curves exist for the $b_0 \neq 0$ case, too, in the form

$$\delta - b_0 = g_{+1}(\varepsilon) . \tag{2.86}$$

In the plane (δ, b_0), these are lines of slope $+1$ (shown as continuous lines in Figure 2.10). Along these D-curves, there exists a characteristic multiplier $\mu = +1$ representing cyclic-fold bifurcation, and equation (2.69) has a periodic solution of period 2π.

If j is odd, that is, $j = 2h + 1, h = 0, 1, \ldots$, then $\lambda = i(h + 1/2)$, and the corresponding characteristic multiplier is

$$\mu = e^{i(h+1/2)2\pi} = e^{i\pi} = -1 . \tag{2.87}$$

In this case, $c_k = \delta + b_0 - (k + h + 1/2)^2$, and $D(\lambda) = 0$ implies the D-curve relation $f_{-1}(\delta + b_0, \varepsilon) = 0$. For the same reason as above, the D-curves exist again in the form

$$\delta + b_0 = g_{-1}(\varepsilon) , \tag{2.88}$$

where $\delta = g_{-1}(\varepsilon)$ gives the U_{-1} D-curves of the classical Mathieu equation. In the parameter plane (δ, b_0), these D-curves are lines of slope -1 (shown as dashed lines in Figure 2.10). Along these D-curves, there exists a characteristic multiplier $\mu = -1$ representing a period-doubling (flip) bifurcation, and equation (2.69) has a nontrivial periodic solution of period 4π.

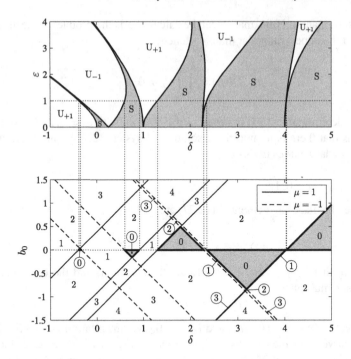

Fig. 2.10 Stability chart with the number of unstable characteristic multipliers for the delayed Mathieu equation (2.69) with $\varepsilon = 1$.

The above analysis showed that the D-curves are straight lines in the plane (δ, b_0) with slopes $+1$, -1 and 0. For varying parameter ε, the lines of slope $+1$ and -1 pass along the stability boundaries of the Ince–Strutt diagram. As mentioned before, these charts are approximate to the same extent as the Ince–Strutt diagram, and the appearance of the delay in the Mathieu equation does not require any further approximation in the stability analysis. The construction of the D-curves in the plane (δ, b_0) for $\varepsilon = 1$ is demonstrated in Figure 2.10. The stability of the domains bounded by the D-curves is determined in the next section.

2.4.3 The Number of Unstable Characteristic Multipliers

The stability of the individual domains separated by the D-curves (2.86) and (2.88) can be determined based on the continuous dependence of the characteristic multipliers on the system parameters. The special case $\varepsilon = 0$ can be treated as a reference: the domains attached to the stable triangles of the Hsu–Bhatt chart (see Figure 2.4) are associated with zero unstable characteristic multipliers. Similarly, some unstable domains can also be identified in this way, and the number of unstable characteristic multipliers can be given based on the case $\varepsilon = 0$. There are, however, some new domains that are not directly connected to any domains of the Hsu–Bhatt chart. The stability of these domains can be determined by the exponent-crossing direction, i.e., by the analysis of the sign of the partial derivative of Re λ with respect to parameter b_0 along the D-curves.

A recursive form for the calculation of the tridiagonal upper left subdeterminants in equation (2.77) can be given as

$$D_{-N} = c_{-N} , \tag{2.89}$$

$$D_{-N+1} = c_{-N}c_{-N+1} - \frac{\varepsilon^2}{4} , \tag{2.90}$$

$$D_k = c_k D_{k-1} - \frac{\varepsilon^2}{4} D_{k-2} , \quad k = -N + 2, \ldots, N . \tag{2.91}$$

Let us denote the partial derivative with respect to b_0 by a prime ($\square' = \partial\square/\partial b_0$) and the substitution of $\lambda = ij/2$ by a hat ($\hat{\square} = \square|_{\lambda=ij/2}$). According to this notation, taking the partial derivatives of expressions (2.74), (2.89), and (2.90) yields

$$c'_k = 2(\lambda + ik)\lambda' - e^{-(\lambda+ik)2\pi} + b_0 2\pi\lambda' e^{-(\lambda+ik)2\pi} , \tag{2.92}$$

$$\hat{c}'_k = \left(2\pi b_0 (-1)^j + i(j + 2k)\right)\lambda' - (-1)^j , \tag{2.93}$$

$$\hat{D}'_{-N} = \left(2\pi b_0 (-1)^j \Gamma_{-N} + i\Omega_{-N}\right)\lambda' - (-1)^j \Gamma_{-N} , \tag{2.94}$$

$$\hat{D}'_{-N+1} = \left(2\pi b_0 (-1)^j \Gamma_{-N+1} + i\Omega_{-N+1}\right)\lambda' - (-1)^j \Gamma_{-N+1} , \tag{2.95}$$

where the coefficients

$$\Gamma_{-N} = 1 ,$$
$$\Omega_{-N} = j - 2N ,$$
$$\Gamma_{-N+1} = \hat{c}_{-N} + \hat{c}_{-N+1} ,$$
$$\Omega_{-N+1} = \hat{c}_{-N}(j - 2N + 2) + \hat{c}_{-N+1}(j - 2N) ,$$

are real numbers, since $\hat{c}_k$ is real for all $k = -N, \ldots, N$. The same differentiation of equation (2.91) yields the recurrence

$$\hat{D}'_k = \hat{c}'_k \hat{D}_{k-1} + \hat{c}_k \hat{D}'_{k-1} - \frac{\varepsilon^2}{4} \hat{D}'_{k-2} , \quad k = -N + 2, \ldots, N . \tag{2.96}$$

It can be proved by induction that (2.96) can be expressed in the same form as (2.94) and (2.95). If, for some k,

$$\hat{D}'_{k-2} = \left(2\pi b_0 (-1)^j \Gamma_{k-2} + i\Omega_{k-2}\right)\lambda' - (-1)^j \Gamma_{k-2} , \tag{2.97}$$

$$\hat{D}'_{k-1} = \left(2\pi b_0 (-1)^j \Gamma_{k-1} + i\Omega_{k-1}\right)\lambda' - (-1)^j \Gamma_{k-1} , \tag{2.98}$$

where $\Gamma_{k-2}, \Gamma_{k-1}, \Omega_{k-2}, \Omega_{k-1}$ are real numbers, then, using (2.93), (2.96), we have

$$\hat{D}'_k = \left(2\pi b_0 (-1)^j \Gamma_k + i\Omega_k\right)\lambda' - (-1)^j \Gamma_k, \quad k = -N+2,\ldots,N, \tag{2.99}$$

where the coefficients

$$\Gamma_k = \hat{D}_{k-1} + \Gamma_{k-1}\hat{c}_k - \frac{\varepsilon^2}{4}\Gamma_{k-2},$$

$$\Omega_k = (j + 2k)\hat{D}_{k-1} + \Omega_{k-1}\hat{c}_k - \frac{\varepsilon^2}{4}\Omega_{k-2},$$

are again real numbers. Together with equations (2.94) and (2.95), this completes the induction.

The final round of recurrence is given by the case $k = N$. Equation $\hat{D}_N = 0$ with $\hat{D}'_N$ defined in (2.99) gives

$$\mathrm{Re}\,\lambda' = \frac{2\pi \Gamma_N^2}{\left(2\pi b_0 (-1)^j \Gamma_N\right)^2 + \Omega_N^2} b_0. \tag{2.100}$$

The exponent-crossing direction along the D-curves is equal to the sign of $\mathrm{Re}\,\lambda'$. Since the coefficient of b_0 in (2.100) is positive, $\mathrm{sgn}(\mathrm{Re}\,\lambda') = \mathrm{sgn}(b_0)$. This means that moving away from the $b_0 = 0$ axis, each D-curve represents characteristic exponents becoming unstable, i.e., crossing the imaginary axis of the complex plane from left to right. The D-curves (2.86) and (2.88) are associated with single critical characteristic multipliers (i.e., $\mu = 1$ and $\mu = -1$, respectively), consequently, the number of unstable characteristic multipliers increases by one at each D-curve as $|b_0|$ is increased. So the only domains of stability are the triangles born from the stable triangles of the $\varepsilon = 0$ case. Since the case $\varepsilon = 0$ is already known (see Figure 2.4), the number of unstable characteristic multipliers can be determined for all the domains by equation (2.100) and by topological considerations. The stability chart can be seen in Figure 2.10 for $\varepsilon = 1$. The stable domains are indicated by gray shading. The frame view of the three-dimensional stability chart in the space $(\delta, b_0, \varepsilon)$ is shown in Figure 2.11.

2.4.4 Damped Case

The stability boundaries for the general damped and delayed Mathieu equation (2.63) with $\tau = 2\pi$ is more complex and they cannot be given in closed form. However, it can be shown that the stability boundaries associated with cyclic-fold and period-doubling (flip) bifurcations remains straight lines of slopes ± 1 in the plane (δ, b_0).

The analysis is very similar to that of the undamped case. Substitution of the trial solution (2.70), Fourier expansion of the periodic terms, application of trigonometric transformations, and balancing the harmonics results in the truncated determinant

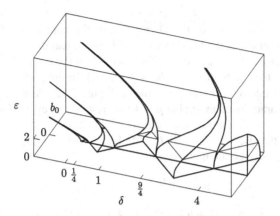

Fig. 2.11 Three-dimensional stability chart of the delayed Mathieu equation (2.69).

$$\tilde{D}(\lambda) = \det \begin{pmatrix} \tilde{c}_{-N} & \varepsilon/2 & & & \\ \varepsilon/2 & \tilde{c}_{-N+1} & \varepsilon/2 & & \\ & \ddots & \ddots & \ddots & \\ & & \varepsilon/2 & \tilde{c}_{N-1} & \varepsilon/2 \\ & & & \varepsilon/2 & \tilde{c}_N \end{pmatrix}, \tag{2.101}$$

where the diagonal elements have the form

$$\tilde{c}_k = \delta + (\lambda + ik)^2 + a_1(\lambda + ik) - b_0 e^{-2\pi(\lambda + ik)} \tag{2.102}$$

instead of c_k defined in equation (2.74).

After the substitution of $\lambda = \pm i\omega$, $\omega \geq 0$, into (2.102), the imaginary part of $\tilde{c}_k$ reads

$$\text{Im}\,\tilde{c}_k = a_1(\omega + k) + b_0 \sin(2\pi\omega) . \tag{2.103}$$

From this point, the proof of the undamped delayed Mathieu equation cannot be continued, since $\text{Im}\,\tilde{c}_k = 0$ does not hold in the cases $\omega = j/2$, $j = 0, 1, \ldots$. This means that D-curves and stability boundaries may exist even for the case $\omega \neq j/2$, $j = 0, 1, \ldots$, and the critical characteristic multipliers can also be complex conjugate pairs of modulus 1. Consequently, secondary Hopf bifurcations may occur in this case, but the corresponding stability boundaries cannot be given in a simple closed form. However, the case $\omega = j/2$, $j = 0, 1, \ldots$, gives

$$\tilde{c}_k = \delta - (k + j/2)^2 - b_0(-1)^j + i(k + j/2)a_1 , \tag{2.104}$$

and the same classification can be done as for the undamped case.

If j is even, that is, $j = 2h$, $h = 0, 1, \ldots$, then $\lambda = ih$, and the corresponding characteristic multiplier is

$$\mu = e^{ih2\pi} = e^{i2\pi} = 1 . \tag{2.105}$$

In this case, $\tilde{c}_k = \delta - b_0 - (k + h)^2 + i(k + h)a_1$, and $\tilde{D}(\lambda) = 0$ gives the relation $\tilde{f}_{+1}(\delta - b_0, \varepsilon, a_1) = 0$ for the D-curves. For the case $b_0 = 0$, the relation $\tilde{f}_{+1}(\delta, \varepsilon, a_1) = 0$ serves as the U_{+1} D-curves of the classical damped Mathieu equation defined in the form $\delta = \tilde{g}_{+1}(\varepsilon, a_1)$. This means that the linear D-curves exist for the $b_0 \neq 0$ case, too, in the form $\delta - b_0 = \tilde{g}_{+1}(\varepsilon, a_1)$. In the plane (δ, b_0), these are lines of slope $+1$ (shown as continuous lines in Figure 2.12). Along these D-curves, there exists a characteristic multiplier $\mu = +1$ representing a cyclic-fold bifurcation, and equation (2.63) has a periodic solution of period 2π.

If j is odd, that is, $j = 2h + 1$, $h = 0, 1, \ldots$, then $\lambda = i(h + 1/2)$, and the corresponding characteristic multiplier is

$$\mu = e^{i(h+1/2)2\pi} = e^{i\pi} = -1 . \tag{2.106}$$

In this case, $\tilde{c}_k = \delta + b_0 - (k + h + 1/2)^2 + i(k + h + 1/2)a_1$, and $\tilde{D}(\lambda) = 0$ implies the relation $\tilde{f}_{-1}(\delta + b_0, \varepsilon, a_1) = 0$ for the D-curves. For the same reason as above, the D-curves exist again in the form $\delta + b_0 = \tilde{g}_{-1}(\varepsilon, a_1)$, where $\delta = \tilde{g}_{-1}(\varepsilon, a_1)$ gives the U_{-1} D-curves of the classical damped Mathieu equation. The D-curves are lines

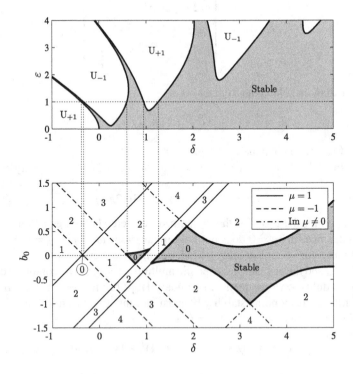

Fig. 2.12 Stability chart with the number of unstable characteristic multipliers for the damped delayed Mathieu equation (2.63) with $\tau = 2\pi$, $a_1 = 0.1$, $\varepsilon = 1$.

of slope -1 in the parameter plane (δ, b_0) (shown as dashed lines in Figure 2.12). Along these D-curves, there exists a characteristic multiplier $\mu = -1$ representing a period-doubling bifurcation, and equation (2.63) has a nontrivial periodic solution of period 4π.

This investigation showed that all the cyclic-fold and period-doubling D-curves are straight lines of slope ± 1 in the plane (δ, b_0). However, in addition to these linear D-curves, other D-curves associated with secondary Hopf bifurcation may also exist, since λ and consequently μ can also be complex at the loss of stability. These D-curves cannot be constructed based on the analysis of the corresponding Hill's deteminant (2.101) in closed form, but they can be determined by numerical techniques. Figure 2.12 presents the stability chart with the number of unstable characteristic multipliers determined by the semi-discretization method. The stable domains are indicated by gray shading. It can be seen that the straight lines obtained by the corresponding $\varepsilon = 1$ section of the Ince–Strutt diagram are indeed stability boundaries. These transition lines are shown as continuous and dashed lines for the $\mu = 1$ and $\mu = -1$ cases, respectively. The D-curves associated with complex characteristic multipliers are shown as dash-dotted lines.

Chapter 3
Semi-discretization

Stability analysis of DDEs with time-periodic coefficients requires the analysis of the eigenvalues of the infinite-dimensional monodromy operator. Generally, stability conditions cannot be given as closed-form functions of the system parameters (the delayed Mathieu equation in Section 2.4 is an exception), but numerical approximations can be used to derive stability properties. Semi-discretization is an efficient numerical method that provides a finite-dimensional matrix approximation of the infinite-dimensional monodromy matrix. This chapter presents the main concept of the semi-discretization method for general linear time-periodic DDEs following [123, 73, 126, 101, 133].

3.1 Basic Ideas

Semi-discretization is a well-known technique used, for example, in the finite element analysis of solid bodies, or in computational fluid mechanics. The basic idea is that the corresponding partial differential equation (PDE) is discretized along the spatial coordinates only, while the time coordinates are unchanged. From a dynamical systems viewpoint, the PDE has an infinite-dimensional state space, which is approximated by the finite-dimensional state space of a high-dimensional ODE.

The same idea can be used for DDEs, but its implementation is not as trivial as for PDEs, since the infinite-dimensional nature of the DDE is due to the presence of the past effects, described by functions embedded also in the time domain, above the past interval $[t - \sigma, t]$, where σ denotes the length of the delay effect (using the notation of (1.13) and (1.15)).

In this section, the main idea of the semi-discretization method is presented for the undamped delayed oscillator

$$\ddot{x}(t) + a_0 x(t) = b_0 x(t - \tau) . \tag{3.1}$$

The stability chart for (3.1) in the parameter plane (a_0, b_0) is presented in Figure 2.4.

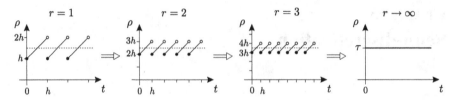

Fig. 3.1 Sampling effect as time-periodic delay for the case $(r + 1/2)h = \tau$.

Introduce first the discrete time scale $t_i = ih$, $i \in \mathbb{Z}$, with h being the discretization step, and consider the DDE

$$\ddot{y}(t) + a_0 y(t) = b_0 y(t_{i-r}) , \qquad t \in [t_i, t_{i+1}) , \tag{3.2}$$

with r being a positive integer. Here, the term $y(t_{i-r})$ refers to the state at $t_{i-r} = (i - r)h$, i.e., the delayed term is constant over the interval $[t_i, t_{i+1})$. This equation for the case $r = 1$ often comes up in control problems modeling the sampling effect with a zero-order hold (see, for example, [162, 1, 15, 209, 225, 258]). Equation (3.2) can also be written in the form

$$\ddot{y}(t) + a_0 y(t) = b_0 y(t - \rho(t)) , \tag{3.3}$$

where

$$\rho(t) = rh - t_i + t , \qquad t \in [t_i, t_{i+1}) , \tag{3.4}$$

is a sawtooth-like time-periodic time delay shown in Figure 3.1. If $r \to \infty$ and $h \to 0$ such that $(r + 1/2)h = \tau$ remains constant, then the time-periodic time delay $\rho(t)$ tends to the constant delay τ, as shown in Figure 3.1. In this sense, (3.2) gives an approximation of the original DDE (3.1) obtained such that the delayed term on the right-hand side of (3.1) is discretized, while all the other (nondelayed) terms on the left-hand side are left in their original form. This is the basic point of the semi-discretization method for delayed systems.

In fact, the sampling effect introduces a periodic parametric excitation at the time delay according to (3.4). Equation (3.2) is therefore a time-periodic DDE (i.e., a DDE with time-periodic point delay) with principal period h. The stability conditions are determined by the eigenvalues of the monodromy operator, which is typically an infinite-dimensional operator for delayed systems. On the other hand, (3.2) can also be considered as a series of ODEs with a piecewise constant forcing on the right-hand side, which implies that a finite-dimensional representation of the monodromy operator can be given as shown below.

For given initial conditions $y_i := y(t_i)$, $\dot{y}_i := \dot{y}(t_i)$ and for a given delayed state variable $y_{i-1} = y(t_{i-1})$, (3.3) can be solved over the discretization interval $[t_i, t_{i+1})$ as an ODE, and the state variable y and its derivative at $t = t_{j+1}$ can be given as

$$y_{i+1} = P_{11}y_i + P_{12}\dot{y}_i + R_1 y_{i-r} , \tag{3.5}$$

$$\dot{y}_{i+1} = P_{21}y_i + P_{22}\dot{y}_i + R_2 y_{i-r} , \tag{3.6}$$

where

$$P_{11} = \frac{\theta_2}{\theta_2 - \theta_1} e^{\theta_1 h} - \frac{\theta_1}{\theta_2 - \theta_1} e^{\theta_2 h} , \tag{3.7}$$

$$P_{12} = \frac{1}{\theta_2 - \theta_1} e^{\theta_2 h} - \frac{1}{\theta_2 - \theta_1} e^{\theta_1 h} , \tag{3.8}$$

$$P_{21} = \frac{\theta_1 \theta_2}{\theta_2 - \theta_1} e^{\theta_1 h} - \frac{\theta_1 \theta_2}{\theta_2 - \theta_1} e^{\theta_2 h} , \tag{3.9}$$

$$P_{22} = \frac{\theta_2}{\theta_2 - \theta_1} e^{\theta_2 h} - \frac{\theta_1}{\theta_2 - \theta_1} e^{\theta_1 h} , \tag{3.10}$$

$$R_1 = \left(1 + \frac{\theta_1}{\theta_2 - \theta_1} e^{\theta_2 h} - \frac{\theta_2}{\theta_2 - \theta_1} e^{\theta_1 h}\right) \frac{b_0}{a_0} , \tag{3.11}$$

$$R_2 = \left(\frac{\theta_1 \theta_2}{\theta_2 - \theta_1} e^{\theta_2 h} - \frac{\theta_1 \theta_2}{\theta_2 - \theta_1} e^{\theta_1 h}\right) \frac{b_0}{a_0} , \tag{3.12}$$

and θ_1 and θ_2 are the roots of the characteristic function

$$D(\theta) = \theta^2 + a_0 \tag{3.13}$$

of the homogeneous part of (3.2), i.e., $\theta_{1,2} = \pm \sqrt{-a_0} = \pm i \sqrt{a_0}$. Equations (3.5) and (3.6) imply the discrete map

$$\underbrace{\begin{pmatrix} y_{i+1} \\ \dot{y}_{i+1} \\ y_i \\ y_{i-1} \\ \vdots \\ y_{i-r+1} \end{pmatrix} = \begin{pmatrix} P_{11} & P_{12} & 0 & 0 & \dots & 0 & R_1 \\ P_{21} & P_{22} & 0 & 0 & \dots & 0 & R_2 \\ 1 & 0 & 0 & 0 & \dots & 0 & 0 \\ 0 & 0 & 1 & 0 & \dots & 0 & 0 \\ \vdots & & & & \ddots & & \vdots \\ 0 & 0 & 0 & 0 & \dots & 1 & 0 \end{pmatrix}}_{:= \mathbf{G}} \begin{pmatrix} y_i \\ \dot{y}_i \\ y_{i-1} \\ y_{i-2} \\ \vdots \\ y_{i-r} \end{pmatrix} . \tag{3.14}$$

The trivial solution of this map, and consequently, the trivial solution of (3.2), are asymptotically stable if all the eigenvalues of the coefficient matrix $\mathbf{G}$ are of modulus less than one (see, e.g., [162, 15, 164, 209]). According to the Floquet theory of time-periodic DDEs, matrix $\mathbf{G}$ is the $(r+2)$-dimensional matrix representation of the monodromy operator of (3.2), and its nonzero eigenvalues are the characteristic multipliers (μ_j, $j = 1, 2, \dots, r+2$). Stability of (3.14) can also be determined by applying the Jury stability criterion to the coefficients of the characteristic polynomial of $\mathbf{G}$, which is a discrete-time analogue of the Routh–Hurwitz stability criterion (see [142, 162]).

Figure 3.2 presents the stability charts in the plane (a_0, b_0) for different values of r such that $(r + 1/2)h = \tau$ is kept with $\tau = 2\pi$. Stable domains are indicated by gray

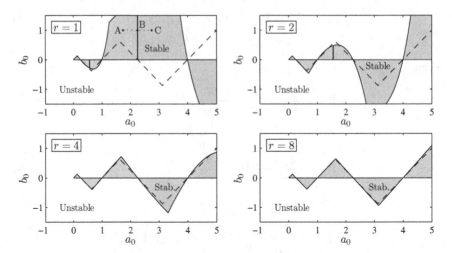

Fig. 3.2 Stability charts for the semi-discrete equation (3.2) for different values of r keeping $(r + 1/2)h = \tau$, $\tau = 2\pi$. The exact stability boundaries of (3.1) are presented by dashed lines.

shading. The diagrams were constructed by point-by-point numerical evaluation of the critical eigenvalues over a 400×200 grid of parameters a_0 and b_0. The exact stability boundaries of (3.1) are also indicated by dashed lines for reference. It can be seen that for increasing r, the stability domains of (3.2) converge to that of (3.1), and for the case $r = 8$, the difference is negligible within the presented parameter limits.

For the case $r = 1$, the stability analysis can be performed in closed form by applying the Jury stability criterion to the characteristic polynomial of $\mathbf{G}$ (that is, in this case, a third-order polynomial). Alternatively, the linear rational complex transformation $\mu = (\nu + 1)/(\nu - 1)$ can be applied to the characteristic polynomial of $\mathbf{G}$, and the polynomial

$$D(\nu) = (\nu - 1)^3 \det\left(\frac{\nu + 1}{\nu - 1}\mathbf{I} - \mathbf{G}\right) \tag{3.15}$$

can be analyzed by the Routh–Hurwitz criterion (for details, see Appendix A.2). After a long but straightforward analysis, the stability conditions can be given as

$$b_0 < a_0 \tag{3.16}$$

and   (3.17)

$$0 < b_0 < \frac{1 + 2\cos(\sqrt{a_0}h)}{1 - \cos(\sqrt{a_0}h)}a_0 \quad \text{or} \quad \frac{1 + 2\cos(\sqrt{a_0}h)}{1 - \cos(\sqrt{a_0}h)}a_0 < b_0 < 0 . \tag{3.18}$$

In addition to the stability boundaries, the vertical lines $a_0 = (k\pi/h)^2$, $k = 0, 1, 2, \ldots$, also represent the appearance of a complex conjugate pair of characteristic multipli-

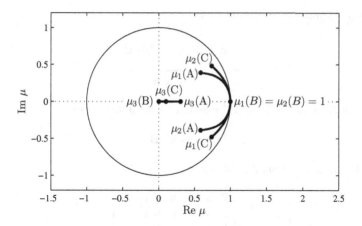

Fig. 3.3 Wandering of the characteristic multipliers of (3.2) with $r = 1$, $h = 4\pi/3$ (giving $\tau = (r + 1/2)h = 2\pi$) along the parameter section ABC in Figure 3.2 (with $b_0 = 1$ and $1.75 \leq a_0 \leq 2.75$).

ers of magnitude equal to 1. These lines, however, do not represent a stability transition, since the corresponding characteristic multipliers just touch the unit circle but do not cross it as the parameter b_0 is increased. Figure 3.3 shows the characteristic multipliers μ_1, μ_2, and μ_3 of the system associated with the parameter section ABC in Figure 3.2 (point B is located just on the vertical line $a_0 = (2\pi/h)^2 = 2.25$). It can be seen that the multipliers μ_1 and μ_2 touch the unit circle but do not cross it.

The semi-discrete approximation for (3.1) can also be applied to the first-order representation

$$\dot{\mathbf{x}}(t) = \mathbf{A}\mathbf{x}(t) + \mathbf{B}\mathbf{u}(t - \tau) , \tag{3.19}$$

$$\mathbf{u}(t) = \mathbf{D}\mathbf{x}(t) , \tag{3.20}$$

with

$$\mathbf{x}(t) = \begin{pmatrix} x(t) \\ \dot{x}(t) \end{pmatrix} , \quad \mathbf{A} = \begin{pmatrix} 0 & 1 \\ -a_0 & 0 \end{pmatrix} , \quad \mathbf{B} = \begin{pmatrix} 0 \\ b_0 \end{pmatrix} , \quad \mathbf{D} = \begin{pmatrix} 1 & 0 \end{pmatrix} . \tag{3.21}$$

Here, $\mathbf{u}(t) = \big(x(t)\big)$ is a 1-dimensional input vector. The corresponding semi-discrete system reads

$$\dot{\mathbf{y}}(t) = \mathbf{A}\mathbf{y}(t) + \mathbf{B}\mathbf{v}(t_{i-r}) , \quad t \in [t_i, t_{i+1}) , \tag{3.22}$$

$$\mathbf{v}(t_i) = \mathbf{D}\mathbf{y}(t_i) . \tag{3.23}$$

For the sake of simplicity, introduce the notation $\mathbf{y}_i := \mathbf{y}(t_i)$ and $\mathbf{v}_i := \mathbf{v}(t_i)$ for all $i \in \mathbb{Z}$. For given initial conditions $\mathbf{y}_i$ and $\mathbf{v}_{i-r}$, the solution at $t = t_{i+1}$ can be given by the variation of constants formula in the form

$$\mathbf{y}_{i+1} = \underbrace{e^{\mathbf{A}h}}_{:= \mathbf{P}} \mathbf{y}_i + \underbrace{\int_0^h e^{\mathbf{A}(h-s)} \, ds \, \mathbf{B} \, \mathbf{v}_{i-r}}_{:= \mathbf{R}} \tag{3.24}$$

(see (A.15) in Appendix A.1). If $\mathbf{A}^{-1}$ exists, then integration gives

$$\mathbf{R} = \left(e^{\mathbf{A}h} - \mathbf{I}\right) \mathbf{A}^{-1} \mathbf{B} . \tag{3.25}$$

Now, (3.24) and (3.23) imply the discrete map

$$\begin{pmatrix} \mathbf{y}_{i+1} \\ \mathbf{v}_i \\ \mathbf{v}_{i-1} \\ \vdots \\ \mathbf{v}_{i-r+1} \end{pmatrix} = \begin{pmatrix} \mathbf{P} & \mathbf{0} & \dots & \mathbf{0} & \mathbf{R} \\ \mathbf{D} & \mathbf{0} & \dots & \mathbf{0} & \mathbf{0} \\ \mathbf{0} & \mathbf{I} & \dots & \mathbf{0} & \mathbf{0} \\ \vdots & & \ddots & & \vdots \\ \mathbf{0} & \mathbf{0} & \dots & \mathbf{I} & \mathbf{0} \end{pmatrix} \begin{pmatrix} \mathbf{y}_i \\ \mathbf{v}_{i-1} \\ \mathbf{v}_{i-2} \\ \vdots \\ \mathbf{v}_{i-r} \end{pmatrix} , \tag{3.26}$$

where $\mathbf{I}$ denotes the 1-dimensional identity matrix. Since (3.22)–(3.23) is the first-order representation of (3.2), the map (3.26) is just identical to (3.14), i.e., the elements $P_{11}, P_{12}, P_{21}, P_{22}$ and R_1, R_2 of $\mathbf{G}$ in (3.14) are in fact the elements of matrices $\mathbf{P}$ and $\mathbf{R}$ in (3.26).

In order to show the efficiency of the semi-discretization method, stability charts for (3.1) are constructed using the full-discretization method for comparison. In the full-discretization process, all the terms in (3.1) are discretized and the derivatives are approximated by finite differences. The approximate (fully discretized) equation reads

$$\frac{y_{i+1} - 2y_i + y_{i-1}}{h^2} + a_0 y_i = b_0 y_{i-r} , \tag{3.27}$$

where the first term is the central difference approximation of the second derivative and r is chosen such that $\tau = (r+1)h$ holds. Equation (3.27) implies the $(r+1) \times (r+1)$ discrete map

$$\underbrace{\begin{pmatrix} y_{i+1} \\ y_i \\ y_{i-1} \\ \vdots \\ y_{i-r+1} \end{pmatrix} = \begin{pmatrix} Q_1 & Q_2 & 0 & \dots & 0 & Q_3 \\ 1 & 0 & 0 & \dots & 0 & 0 \\ 0 & 1 & 0 & \dots & 0 & 0 \\ \vdots & & & \ddots & & \vdots \\ 0 & 0 & 0 & \dots & 1 & 0 \end{pmatrix} \begin{pmatrix} y_i \\ y_{i-1} \\ y_{i-2} \\ \vdots \\ y_{i-r} \end{pmatrix}}_{:= \mathbf{H}} , \tag{3.28}$$

where $Q_1 = 2 - a_0 h^2$, $Q_2 = -1$ and $Q_3 = b_0 h^2$. The stability of the approximate system is determined by the eigenvalues of the coefficient matrix $\mathbf{H}$. The stability charts constructed by the above full-discretization method are shown in Figure 3.4. Stable domains are shown in gray, and the exact stability boundaries of (3.1) are also represented by dashed lines. Comparison to Figure 3.2 shows that the stability

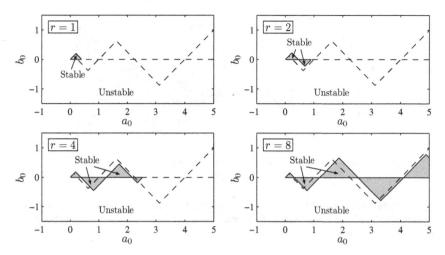

Fig. 3.4 Stability charts for (3.27) obtained by the central difference approximation of the second derivative in (3.1) for different values of r keeping $(r + 1)h = \tau$, $\tau = 2\pi$. The exact stability boundaries of (3.1) are represented by dashed lines.

boundaries obtained by the semi-discretization method converge faster to the exact boundaries than those obtained by full-discretization.

Note that if other than central difference is used to approximate the second derivative, then the convergence is even worse. If the backward difference is used to approximate the second derivative of the state, then the fully discretized system reads

$$\frac{y_i - 2y_{i-1} + y_{i-2}}{h^2} + a_0 y_i = b_0 y_{i-r} \qquad (3.29)$$

with $\tau = rh$. For the forward difference approximation, the resulting fully discretized system is

$$\frac{y_{i+2} - 2y_{i+1} + y_i}{h^2} + a_0 y_i = b_0 y_{i-r} \qquad (3.30)$$

with $\tau = (r + 2)h$. Similarly to (3.28), these equations can be transformed into an $r \times r$ and an $(r + 2) \times (r + 2)$ discrete map, respectively. The corresponding stability boundaries for (3.29) and (3.30) are shown in Figures 3.5 and 3.6, respectively. Stable domains associated with different values of r are represented by different shades of gray. It can be seen that the convergence of the full-discretization method using either backward or forward difference approximation of the second derivative is much worse than the convergence obtained by the central difference approximation shown in Figure 3.4.

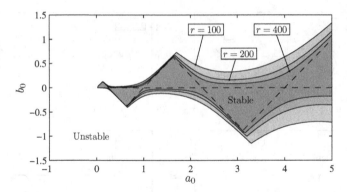

Fig. 3.5 Stability charts for (3.29) obtained by backward difference approximation of the second derivative in (3.1) for different values of r keeping $rh = \tau$, $\tau = 2\pi$. The exact stability boundaries of (3.1) are represented by dashed lines.

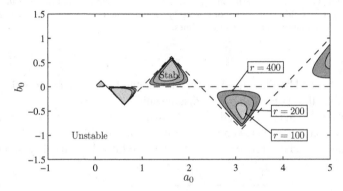

Fig. 3.6 Stability charts for (3.30) obtained by forward difference approximation of the second derivative in (3.1) for different values of r keeping $(r + 2)h = \tau$, $\tau = 2\pi$. The exact stability boundaries of (3.1) are represented by dashed lines.

3.2 General Concept

The semi-discretization method is applied to time-periodic DDEs of the form

$$\dot{\mathbf{x}}(t) = \mathbf{A}(t)\mathbf{x}(t) + \int_{-\sigma}^{0} \mathbf{K}(\vartheta, t)\mathbf{u}(t + \vartheta)\,\mathrm{d}\vartheta\,, \qquad (3.31)$$

$$\mathbf{u}(t) = \mathbf{D}\mathbf{x}(t)\,, \qquad (3.32)$$

where $\mathbf{x}(t) \in \mathbb{R}^n$ is the state, $\mathbf{u}(t) \in \mathbb{R}^m$ is the input, $\mathbf{A}(t + T) = \mathbf{A}(t)$ and $\mathbf{D}$ are $n \times n$ and $m \times n$ matrices, respectively, and $\mathbf{K}(\vartheta, t + T) = \mathbf{K}(\vartheta, t)$ is a measurable time-periodic $n \times m$ kernel function matrix that may comprise a measurable distribution and finitely many shifted Dirac delta distributions. Thus, the kernel $\mathbf{K}(\vartheta, t)$ can be written as

$$\mathbf{K}(\vartheta, t) = \mathbf{W}(\vartheta, t) + \sum_{j=1}^{g} \mathbf{B}_j(t)\delta(\vartheta + \tau_j(t)) , \qquad (3.33)$$

where $\mathbf{W}(\vartheta, t + T) = \mathbf{W}(\vartheta, t)$ is an $n \times m$ measurable function, $\mathbf{B}_j(t + T) = \mathbf{B}_j(t)$ are $n \times m$ matrices, $\delta(\vartheta)$ denotes the Dirac delta distribution, $\tau_j(t + T) = \tau_j(t) \geq 0$ for all j, and $g \in \mathbb{N}$. Thus, system (3.31)–(3.32) can also be written as

$$\dot{\mathbf{x}}(t) = \mathbf{A}(t)\mathbf{x}(t) + \int_{-\sigma}^{0} \mathbf{W}(\vartheta, t)\mathbf{u}(t + \vartheta) \, d\vartheta + \sum_{j=1}^{g} \mathbf{B}_j(t)\mathbf{u}(t - \tau_j(t)) , \qquad (3.34)$$

$$\mathbf{u}(t) = \mathbf{D}\mathbf{x}(t) , \qquad (3.35)$$

or

$$\dot{\mathbf{x}}(t) = \mathbf{A}(t)\mathbf{x}(t) + \int_{-\sigma}^{0} \mathbf{W}(\vartheta, t)\mathbf{D}\mathbf{x}(t + \vartheta) \, d\vartheta + \sum_{j=1}^{g} \mathbf{B}_j(t)\mathbf{D}\mathbf{x}(t - \tau_j(t)) . \qquad (3.36)$$

Semi-discretization is based on the discrete time scale $t_i = ih$, $i \in \mathbb{Z}$, where the discretization step is $h = T/p$ with T being the principal period of the system and p an integer approximation parameter. Semi-discretization of the general system (3.34)–(3.35) consists of three different steps of approximations.

1. First, the distributed delay term is approximated by a linear combination of point delays. The point of this step is that the weight function in (3.34) is approximated by a linear combination of shifted Dirac delta distributions in the form

$$\mathbf{W}(\vartheta, t) \approx \tilde{\mathbf{W}}(\vartheta, t) = \sum_{k=1}^{f} \tilde{\mathbf{W}}_k(t) \, \delta\left(\vartheta + \left(k - \tfrac{1}{2}\right)h\right) , \qquad (3.37)$$

where

$$\tilde{\mathbf{W}}_k(t) = \int_{-kh}^{-(k-1)h} \mathbf{W}(\vartheta, t) \, d\vartheta , \qquad k = 1, 2, \ldots, f - 1 , \qquad (3.38)$$

$$\tilde{\mathbf{W}}_f(t) = \int_{-\sigma}^{-(f-1)h} \mathbf{W}(\vartheta, t) \, d\vartheta , \qquad (3.39)$$

$f = \text{ceil}(\sigma/h)$, with ceil being the ceiling function (i.e., $\text{ceil}(x)$ is the smallest integer not less than x) and $\delta(\vartheta)$ denotes the Dirac delta distribution. The sketch of this approximation step is shown in Figure 3.7. Using (3.37), the distributed delay term in (3.34) is approximated as

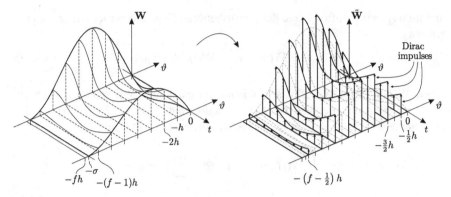

Fig. 3.7 Approximation of the weight function of the distributed delay by shifted Dirac delta distributions.

$$\int_{-\sigma}^{0} \mathbf{W}(\vartheta, t)\,\mathbf{u}(t + \vartheta)\,\mathrm{d}\vartheta \approx \int_{-fh}^{0} \tilde{\mathbf{W}}(\vartheta, t)\,\mathbf{u}(t + \vartheta)\,\mathrm{d}\vartheta$$

$$= \sum_{k=1}^{f} \tilde{\mathbf{W}}_k(t)\,\mathbf{u}\left(t - \left(k - \tfrac{1}{2}\right)h\right) = \sum_{k=1}^{f} \tilde{\mathbf{W}}_k(t)\,\mathbf{u}(t - \tilde{\tau}_k) \quad (3.40)$$

with $\tilde{\tau}_k = \left(k - \tfrac{1}{2}\right)h$, $k = 1, 2, \ldots, f$. In this step, the original system (3.34)–(3.35) is approximated by

$$\dot{\tilde{\mathbf{x}}}(t) = \mathbf{A}(t)\tilde{\mathbf{x}}(t) + \sum_{j=1}^{g+f} \mathbf{B}_j(t)\tilde{\mathbf{u}}(t - \tau_j(t)) , \qquad (3.41)$$

$$\tilde{\mathbf{u}}(t) = \mathbf{D}\tilde{\mathbf{x}}(t) , \qquad (3.42)$$

such that $\mathbf{B}_{g+k}(t) = \tilde{\mathbf{W}}_k(t)$ and $\tau_{g+k}(t) = \tilde{\tau}_k$ for $k = 1, 2, \ldots, f$. For discretization step h small enough, the solution $\tilde{\mathbf{x}}(t)$ of (3.41)–(3.42) tends to the solution $\mathbf{x}(t)$ of (3.34)–(3.35).

2. Second, the time-dependent matrices $\mathbf{A}(t)$, $\mathbf{B}_j(t)$ and the time-dependent delays $\tau_j(t)$ are approximated by their averages

$$\mathbf{A}_i = \frac{1}{h} \int_{t_i}^{t_{i+1}} \mathbf{A}(t)\,\mathrm{d}t , \qquad (3.43)$$

$$\mathbf{B}_{j,i} = \frac{1}{h} \int_{t_i}^{t_{i+1}} \mathbf{B}_j(t)\,\mathrm{d}t , \qquad j = 1, 2, \ldots, g + f , \qquad (3.44)$$

$$\tau_{j,i} = \frac{1}{h} \int_{t_i}^{t_{i+1}} \tau_j(t)\,\mathrm{d}t , \qquad j = 1, 2, \ldots, g + f , \qquad (3.45)$$

in each discretization interval $[t_i, t_{i+1}]$, $i = 1, 2, \ldots, p$. In this step, (3.41)–(3.42) is approximated by a DDE of constant coefficients with constant point delays in the form

$$\dot{\tilde{\mathbf{x}}}(t) = \mathbf{A}_i \tilde{\mathbf{x}}(t) + \sum_{j=1}^{g+f} \mathbf{B}_{j,i} \, \tilde{\mathbf{u}}(t - \tau_{j,i}) \,, \qquad t \in [t_i, t_{i+1}) \,, \tag{3.46}$$

$$\tilde{\mathbf{u}}(t) = \mathbf{D} \, \tilde{\mathbf{x}}(t) \,. \tag{3.47}$$

This step is basically equivalent to the piecewise autonomous approximation of nonautonomous systems, similarly to (1.9). The solution $\tilde{\mathbf{x}}(t)$ of (3.46)–(3.47) tends to the solution $\tilde{\mathbf{x}}(t)$ of (3.41)–(3.42) if the discretization step h tends to zero. Note that the order of approximation steps 1 and 2 can be reversed.

3. Third, in each discretization interval $[t_i, t_{i+1})$, the delayed terms $\mathbf{u}(t - \tau_{j,i})$ are approximated by constants

$$\mathbf{u}(t - \tau_{j,i}) \approx \mathbf{u}(t_{i-r_{j,i}}) = \mathbf{u}((i - r_{j,i})h) \,, \tag{3.48}$$

for $j = 1, 2, \ldots, g + f$ and $i = 1, 2, \ldots, p$, where

$$r_{j,i} = \mathrm{int}\left(\tau_{j,i}/h\right) \,, \tag{3.49}$$

with int denoting the integer-part function. This step corresponds to the sawtooth-like time-periodic approximation of the delayed term shown in Figure 3.8 (see also earlier in Figure 3.1) in the form

$$\tau_{j,i} \approx \rho_{j,i}(t) = r_{j,i}h - t_i + t \,, \qquad t \in [t_i, t_{i+1}) \,. \tag{3.50}$$

In this step, (3.46)–(3.47) is approximated by a series of ODEs of constant coefficients in the form

$$\dot{\mathbf{y}}(t) = \mathbf{A}_i \mathbf{y}(t) + \sum_{j=1}^{g+f} \mathbf{B}_{j,i} \mathbf{v}(t_{i-r_{j,i}}) \,, \qquad t \in [t_i, t_{i+1}) \,, \tag{3.51}$$

$$\mathbf{v}(t_i) = \mathbf{D}\mathbf{y}(t_i) \,. \tag{3.52}$$

If the discretization step h tends to 0, then the solution $\mathbf{y}(t)$ of (3.51)–(3.52) tends to the solution $\tilde{\mathbf{x}}(t)$ of (3.46)–(3.47). Note that linear or higher-order polynomial approximation of the delayed terms can also be applied instead of the constant one. Such higher-order approximations are shown in Section 3.3.

For the sake of simplicity, use the notation $\mathbf{y}_i := \mathbf{y}(t_i)$ and $\mathbf{v}_i := \mathbf{v}(t_i)$ for all $i \in \mathbb{Z}$. By means of the above three steps of approximations, the original DDE (3.34)–(3.35) is approximated by the ODE (3.51)–(3.52) with piecewise constant coefficients. If the initial conditions $\mathbf{y}_i$ and $\mathbf{v}_{i-r_{j,i}}$ are given (for all $j = 1, 2, \ldots, g+f$), then the solution of (3.51)–(3.52) over one discrete step can be formulated as

$$\mathbf{y}_{i+1} = \mathbf{P}_i \mathbf{y}_i + \sum_{j=1}^{g+f} \mathbf{R}_{j,i} \mathbf{v}_{i-r_{j,i}} \,, \tag{3.53}$$

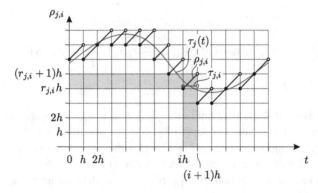

Fig. 3.8 Approximation of the time-varying delay $\tau_j(t)$ by the piecewise linear one $\rho_{j,i}$.

where

$$\mathbf{P}_i = \mathrm{e}^{\mathbf{A}_i h}, \tag{3.54}$$

$$\mathbf{R}_{j,i} = \int_0^h \mathrm{e}^{\mathbf{A}_i(h-s)}\, \mathrm{d}s\, \mathbf{B}_{j,i} \tag{3.55}$$

(see the variation of constants formula (A.15) in Appendix A.1). If $\mathbf{A}_i^{-1}$ exists, then integration gives

$$\mathbf{R}_{j,i} = \left(\mathrm{e}^{\mathbf{A}_i h} - \mathbf{I}\right)\mathbf{A}_i^{-1}\mathbf{B}_{j,i}, \tag{3.56}$$

where $\mathbf{I}$ is the $n \times n$ identity matrix. Equations (3.53) and (3.52) imply the discrete map

$$\mathbf{z}_{i+1} = \mathbf{G}_i \mathbf{z}_i, \tag{3.57}$$

where

$$\mathbf{z}_i = \left(\mathbf{y}_i \ \ \mathbf{v}_{i-1} \ \ \mathbf{v}_{i-2} \ \ \dots \ \ \mathbf{v}_{i-r}\right)^T \tag{3.58}$$

is an augmented state vector and the coefficient matrix reads

$$\mathbf{G}_i = \begin{pmatrix} \mathbf{P}_i & \mathbf{0} & \cdots & \mathbf{0} & \mathbf{0} \\ \hline \mathbf{D} & \mathbf{0} & \cdots & \mathbf{0} & \mathbf{0} \\ \mathbf{0} & \mathbf{I} & \cdots & \mathbf{0} & \mathbf{0} \\ \vdots & & \ddots & & \vdots \\ \mathbf{0} & \mathbf{0} & \cdots & \mathbf{I} & \mathbf{0} \end{pmatrix} + \sum_{j=1}^{g+f} \begin{pmatrix} \mathbf{0} & \mathbf{0} & \cdots & \mathbf{0} & \mathbf{R}_{j,i} & \mathbf{0} & \cdots & \mathbf{0} \\ \hline \mathbf{0} & \mathbf{0} & \cdots & \mathbf{0} & \mathbf{0} & \mathbf{0} & \cdots & \mathbf{0} \\ \mathbf{0} & \mathbf{0} & \cdots & \mathbf{0} & \mathbf{0} & \mathbf{0} & \cdots & \mathbf{0} \\ \vdots & \vdots & & \vdots & \vdots & \vdots & & \vdots \\ \mathbf{0} & \mathbf{0} & \cdots & \mathbf{0} & \mathbf{0} & \mathbf{0} & \cdots & \mathbf{0} \end{pmatrix}. \tag{3.59}$$

This is an $(n + rm) \times (n + rm)$ matrix with $r = \max(r_{j,i})$, $j = 1, 2, \dots, g + f$, $i = 1, 2, \dots, p$, which can be divided into four blocks, as shown by the lines in (3.59). The left upper block is the $n \times n$ matrix $\mathbf{P}_i$. The right upper block consists of r pieces of $n \times m$ matrices numbered in (3.59). Matrices $\mathbf{R}_{j,i}$ are located at the $r_{j,i}$th place within this block. The left lower block of $\mathbf{G}_i$ consists of r pieces of $m \times n$ matrices (matrix $\mathbf{D}$ and zero matrices). The right lower block is an $rm \times rm$ block containing zero and identity matrices $\mathbf{I}$ of size $m \times m$.

Recalling that $T = ph$, p repeated applications of (3.57) with initial state $\mathbf{z}_0$ gives the monodromy mapping

$$\mathbf{z}_p = \mathbf{\Phi}\mathbf{z}_0, \tag{3.60}$$

where

$$\mathbf{\Phi} = \mathbf{G}_{p-1}\mathbf{G}_{p-2} \cdots \mathbf{G}_0 \tag{3.61}$$

is an $(n + rm)$-dimensional matrix representation of the monodromy operator of (3.51)–(3.52). In this way, $\mathbf{\Phi}$ provides a finite-dimensional approximation of the infinite-dimensional monodromy operator of the original system (3.34)–(3.35).

The stability of the approximate system (3.51)–(3.52) can be assessed by the eigenvalue analysis of matrix $\mathbf{\Phi}$. If all the eigenvalues are inside the unit circle of the complex plane, then the system (3.51)–(3.52) is asymptotically stable. Since semi-discretization preserves asymptotic stability of the original system (3.34)–(3.35), as shown in [101] (and, in general, discretization techniques preserve asymptotic stability for DDEs; see [95]), the method can be used to construct approximate stability charts.

The approximation parameter p is related to the resolution of the principal period such that $T = ph$; therefore, the integer p is called the *period resolution*. The integer r is related to the discretization of the state $\mathbf{x}_t$ over the delay interval $[-\sigma, 0]$ such that $\sigma \approx rh$. Therefore, the integer r is called the *delay resolution* (note that σ is the maximum delay in the system). The number of matrices $\mathbf{G}_i$ to be multiplied to obtain the monodromy matrix $\mathbf{\Phi}$ in (3.61) is equal to the period resolution p. The size of the matrices $\mathbf{G}_i$ (and the size of the monodromy matrix $\mathbf{\Phi}$) is equal to $(n + rm)$. Recall that $r = \max(r_{j,i})$, $j = 1, 2, \ldots, g + f$, $i = 1, 2, \ldots, p$, and $r_{j,i} = \text{int}(\tau_{j,i}/h) = \text{int}(p\tau_{j,i}/T)$. Here, $r_{j,i}$ is the *particular delay resolution* associated with the particular delay $\tau_{j,i}$. Thus, the larger the period resolution p, the larger the size of the approximate monodromy matrix. If p tends to infinity, then $\mathbf{\Phi}$ converges to the infinite-dimensional monodromy operator of the original system (3.34)–(3.35).

The convergence of the semi-discretization method can be visualized by plotting the characteristic multipliers in the complex plane. Let us denote the characteristic multipliers of the original system (3.34)–(3.35) by μ_k, $k = 1, 2, \ldots$, and the characteristic multipliers of the approximate semi-discrete system (3.51)–(3.52) by $\tilde{\mu}_k$, $k = 1, 2, \ldots, (n + rm)$. Let the circles of center $\tilde{\mu}_k$ and radius ε be denoted by $S_{\tilde{\mu}_k, \varepsilon}$. For any small $\varepsilon > 0$, there exists an integer $M(\varepsilon)$ such that for every $p > M(\varepsilon)$, the set $\bigcup_{k=1}^{n+rm} S_{\tilde{\mu}_k, \varepsilon}$ contains exactly $n + rm$ characteristic multipliers μ_k of (3.34)–(3.35), and all the other characteristic multipliers have modulus less than ε.

Thus, if all the characteristic multipliers of (3.51)–(3.52) have modulus less than 1, then by choosing $\varepsilon = \frac{1}{2}(1 - \max_j |\tilde{\mu}_j|)$, the finite approximation number $M(\varepsilon)$ exists, and if $p > M(\varepsilon)$, then the discretized system and the original system have the same stability properties (see Figure 3.9 with $n + rm = 5$).

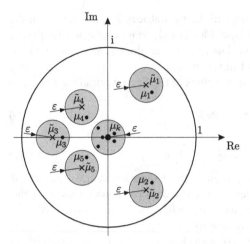

Fig. 3.9 Locations of the exact (μ) and the approximate ($\tilde{\mu}$) characteristic multipliers.

During the construction of stability charts, the numerical efficiency of the method depends, of course, on the choice of the period resolution p. A lower estimate can be given based on the above derivation of $M(\varepsilon)$, but in practice, it is easier to use a trial-and-error technique and to check how the stability boundaries converge for increasing p. The value of the required p depends on the system's parameters, which means that the construction of the stability charts can be optimized by choosing different values in different domains of the charts. This will be explained by the examples in Section 4.1.

3.3 Toward Higher-Order Methods

In the previous section, the semi-discretization method was presented for general systems including distributed delays such that the delayed terms were approximated by piecewise constant ones over each discretization step. In this section, higher-order methods are presented, where the delayed terms are approximated by higher-order polynomials of time t. Here, for the sake of simplicity, the method is presented for systems with point delays. Systems with distributed delays should first be approximated by systems with point delays according to approximation step 1 in Section 3.2; then the higher-order approximation can be applied, as explained below.

3.3.1 Higher-Order Approximations

Consider the time-periodic DDE with multiple time-periodic point delays of the form

$$\dot{\mathbf{x}}(t) = \mathbf{A}(t)\mathbf{x}(t) + \sum_{j=1}^{g} \mathbf{B}_j(t)\mathbf{u}(t - \tau_j(t)) , \qquad (3.62)$$

$$\mathbf{u}(t) = \mathbf{D}\mathbf{x}(t) , \qquad (3.63)$$

where $\mathbf{x}(t) \in \mathbb{R}^n$ is the state, $\mathbf{u}(t) \in \mathbb{R}^m$ is the input, $\mathbf{A}(t + T) = \mathbf{A}(t)$ and $\mathbf{B}_j(t + T) = \mathbf{B}_j(t)$, $j = 1, 2, \ldots, g$, are $n \times n$ and $n \times m$ time-periodic matrices, respectively, $\mathbf{D}$ is an $m \times n$ constant matrix, and $\tau_j(t + T) = \tau_j(t) > 0$, $j = 1, 2, \ldots, g$. The principal period of the system is T. Note that (3.62)–(3.63) can also be written in the form

$$\dot{\mathbf{x}}(t) = \mathbf{A}(t)\mathbf{x}(t) + \sum_{j=1}^{g} \mathbf{B}_j(t)\mathbf{D}\mathbf{x}(t - \tau_j(t)) . \qquad (3.64)$$

The main point of higher-order semi-discretization methods is that the time-periodic coefficients are approximated by piecewise constant ones, and the delayed terms are approximated by linear combinations of some discrete delayed values of the state variable $\mathbf{x}$, while the nondelayed terms are left in their original form. Consider the discrete time scale $t_i = ih$, $i \in \mathbb{Z}$, such that the time step is $h = T/p$ with $p \in \mathbb{Z}^+$ being the period resolution. The approximating semi-discrete system is formulated as

$$\dot{\mathbf{y}}(t) = \mathbf{A}_i\mathbf{y}(t) + \sum_{j=1}^{g} \mathbf{B}_{j,i}\boldsymbol{\Gamma}_{j,i}^{(q)}(t - \tau_{j,i}) , \quad t \in [t_i, t_{i+1}) , \qquad (3.65)$$

$$\boldsymbol{\Gamma}_{j,i}^{(q)}(t - \tau_{j,i}) = \sum_{k=0}^{q} \left(\prod_{l=0, l \neq k}^{q} \frac{t - \tau_{j,i} - (i + l - r_{j,i})h}{(k - l)h} \right) \mathbf{v}(t_{i+k-r_{j,i}}) , \qquad (3.66)$$

$$\mathbf{v}(t_i) = \mathbf{D}\mathbf{y}(t_i), \qquad (3.67)$$

where

$$\mathbf{A}_i = \frac{1}{h} \int_{t_i}^{t_{i+1}} \mathbf{A}(t) \, dt , \qquad (3.68)$$

$$\mathbf{B}_{j,i} = \frac{1}{h} \int_{t_i}^{t_{i+1}} \mathbf{B}_j(t) \, dt , \qquad j = 1, 2, \ldots, g , \qquad (3.69)$$

$$\tau_{j,i} = \frac{1}{h} \int_{t_i}^{t_{i+1}} \tau_j(t) \, dt , \qquad j = 1, 2, \ldots, g , \qquad (3.70)$$

are the piecewise constant approximations of $\mathbf{A}(t)$, $\mathbf{B}_j(t)$, and $\tau_j(t)$, $j = 1, 2, \ldots, g$, over the discretization interval $[t_i, t_{i+1})$. Use again the notation $\mathbf{y}_i := \mathbf{y}(t_i)$ and $\mathbf{v}_i := \mathbf{v}(t_i)$ for all $i \in \mathbb{Z}$. The delayed term $\boldsymbol{\Gamma}_{j,i}^{(q)}(t - \tau_{j,i})$ is a qth-order Lagrange

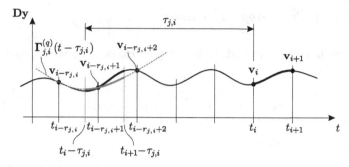

Fig. 3.10 Approximation of the delayed term $\mathbf{D}y(t - \tau_{j,i})$ by the polynomial $\mathbf{\Gamma}_{j,i}^{(q)}(t - \tau_{j,i})$, shown by a dashed line (in the depicted case, $q = 2$).

polynomial interpolation of $\mathbf{D}(t)\mathbf{y}(t)$ in $t \in [t_{i-r_{j,i}}, t_{i-r_{j,i}+q}]$ using the discrete values $\mathbf{v}_{i-r_{j,i}}, \mathbf{v}_{i-r_{j,i}+1}, \ldots, \mathbf{v}_{i-r_{j,i}+q}$. The particular delay resolution $r_{j,i}$ is defined by

$$r_{j,i} = \text{int}\left(\frac{\tau_{j,i}}{h} + \frac{q}{2}\right), \qquad j = 1, 2, \ldots, g, \qquad i = 1, 2, \ldots, p, \qquad (3.71)$$

where int denotes the integer-part function. Thus, we have two approximation parameters: the period resolution p, which is the number of discrete steps over the principal period T, and the order q of the approximation of the delayed term. Note that the particular delay resolution $r_{j,i}$ is defined such that $(r_{j,i} - q/2)h \approx \tau_{j,i}$. The concept of the approximation scheme is illustrated in Figure 3.10, where the dashed curve indicates the approximating polynomial $\mathbf{\Gamma}_{j,i}^{(q)}(t - \tau_{j,i})$.

The key feature of the semi-discretization method is that the approximate system (3.65)–(3.67) can be solved analytically over the discretization interval $t \in [t_i, t_{i+1})$ for given initial values $\mathbf{y}_i$ and $\mathbf{v}_{i+k-r_{j,i}}$, $k = 0, 1, \ldots, q$, $j = 1, 2, \ldots, g$, in the form

$$\mathbf{y}_{i+1} = \mathbf{P}_i \mathbf{y}_i + \sum_{j=1}^{g} \sum_{k=0}^{q} \mathbf{R}_{j,i,k} \mathbf{v}_{i+k-r_{j,i}}, \qquad (3.72)$$

where

$$\mathbf{P}_i = e^{\mathbf{A}_i h}, \qquad (3.73)$$

$$\mathbf{R}_{j,i,k} = \int_{t_i}^{t_{i+1}} e^{\mathbf{A}_i(t_{i+1}-s)} \left(\prod_{l=0,\, l\neq k}^{q} \frac{s - \tau_{j,i} - (i + l - r_{j,i})h}{(k-l)h} \right) \mathbf{B}_{j,i}\, ds$$

$$= \int_0^h e^{\mathbf{A}_i(h-s)} \left(\prod_{l=0,\, l\neq k}^{q} \frac{s - \tau_{j,i} - (l - r_{j,i})h}{(k-l)h} \right) \mathbf{B}_{j,i}\, ds \qquad (3.74)$$

(see the variation of constants formula (A.15) in Appendix A.1). From this point, the steps are the same presented in Section 3.2. Equations (3.72) and (3.67) imply the discrete map

$$\mathbf{z}_{i+1} = \mathbf{G}_i \mathbf{z}_i, \tag{3.75}$$

where

$$\mathbf{z}_i = \begin{pmatrix} \mathbf{y}_i & \mathbf{v}_{i-1} & \mathbf{v}_{i-2} & \cdots & \mathbf{v}_{i-r} \end{pmatrix}^T \tag{3.76}$$

is an augmented state vector and the coefficient matrix reads

$$\mathbf{G}_i = \begin{pmatrix} \mathbf{P}_i & \mathbf{0} & \cdots & \mathbf{0} & \mathbf{0} \\ \hline \mathbf{D} & \mathbf{0} & \cdots & \mathbf{0} & \mathbf{0} \\ \mathbf{0} & \mathbf{I} & \cdots & \mathbf{0} & \mathbf{0} \\ \vdots & & \ddots & & \vdots \\ \mathbf{0} & \mathbf{0} & \cdots & \mathbf{I} & \mathbf{0} \end{pmatrix} + \sum_{j=1}^{g} \begin{pmatrix} \overset{1}{\mathbf{0}} & \mathbf{0} & \cdots & \mathbf{0} & \overset{r_{j,i}-q}{\mathbf{R}_{j,i,q}} & \cdots & \overset{r_{j,i}}{\mathbf{R}_{j,i,0}} & \mathbf{0} & \cdots & \overset{r}{\mathbf{0}} \\ \hline \mathbf{0} & \mathbf{0} & \cdots & \mathbf{0} & \mathbf{0} & \cdots & \mathbf{0} & \mathbf{0} & \cdots & \mathbf{0} \\ \mathbf{0} & \mathbf{0} & \cdots & \mathbf{0} & \mathbf{0} & \cdots & \mathbf{0} & \mathbf{0} & \cdots & \mathbf{0} \\ \vdots & \vdots & & \vdots & \vdots & & \vdots & \vdots & & \vdots \\ \mathbf{0} & \mathbf{0} & \cdots & \mathbf{0} & \mathbf{0} & \cdots & \mathbf{0} & \mathbf{0} & \cdots & \mathbf{0} \end{pmatrix}. \tag{3.77}$$

This is an $(n + rm) \times (n + rm)$ matrix, where $r = \max(r_{j,i})$ is the delay resolution such that $\tau_{\max} \approx (r - q/2)h$ with $\tau_{\max} = \max(\tau_{j,i})$. Matrix $\mathbf{G}_i$ can be divided into four blocks, as shown by the lines in (3.77). The left upper block is the $n \times n$ matrix $\mathbf{P}_i$. The right upper block consists of r pieces of $n \times m$ matrices numbered in (3.77). Matrices $\mathbf{R}_{j,i,k}$, $k = 0, 1, \ldots, q$, are located at the $(r_{j,i} - k)$th place within this block. The left lower block of $\mathbf{G}_i$ consists of r pieces of $m \times n$ matrices ($\mathbf{D}$ and zero matrices). The right lower block is an $rm \times rm$ block containing zero matrices and identity matrices $\mathbf{I}$ of size $m \times m$.

Utilizing that $T = ph$, p repeated applications of (3.75) with initial state $\mathbf{z}_0$ gives the monodromy mapping

$$\mathbf{z}_p = \mathbf{\Phi} \mathbf{z}_0, \tag{3.78}$$

where

$$\mathbf{\Phi} = \mathbf{G}_{p-1} \mathbf{G}_{p-2} \cdots \mathbf{G}_0 \tag{3.79}$$

is an $(n + rm)$-dimensional matrix representation of the monodromy operator of (3.65)–(3.67), which is at the same time a finite-dimensional approximation of the infinite-dimensional monodromy operator of the original system (3.62)–(3.63). If all the eigenvalues of $\mathbf{\Phi}$ are inside the unit circle of the complex plane, then the approximate system (3.65)–(3.67) is asymptotically stable. Since discretization techniques preserve asymptotic stability for DDEs (see [95, 101]), the stability charts of the approximate system (3.65)–(3.67) give an approximation for the stability charts of the original time-periodic DDE (3.62)–(3.63).

The above formulas give the steps of the semi-discretization method for arbitrary approximation order q. For the sake of completeness, some special cases of these formulas are presented for the zeroth- and the first-order approximations in the next subsections.

Zeroth-Order Semi-Discretization

For the zeroth-order case ($q = 0$), (3.66) and (3.71) give $\boldsymbol{\Gamma}_{j,i}^{(q)}(t - \tau_{j,i}) \equiv \mathbf{v}_{i-r_{j,i}}$ and $r_{j,i} = \text{int}(\tau_{j,i}/h)$, and the approximate system reads

$$\dot{\mathbf{y}}(t) = \mathbf{A}_i \mathbf{y}(t) + \sum_{j=1}^{g} \mathbf{B}_{j,i} \mathbf{v}(t_{i-r_{j,i}}) , \quad t \in [t_i, t_{i+1}) , \tag{3.80}$$

$$\mathbf{v}(t_i) = \mathbf{D}\mathbf{y}(t_i) . \tag{3.81}$$

This corresponds to a piecewise constant approximation of the delayed term according to Figure 3.11.

Using the notation $\mathbf{y}_i := \mathbf{y}(t_i)$ and $\mathbf{v}_i := \mathbf{v}(t_i)$, $i \in \mathbb{Z}$, the solution over one discrete step can be formulated as

$$\mathbf{y}_{i+1} = \mathbf{P}_i \mathbf{y}_i + \sum_{j=1}^{g} \mathbf{R}_{j,i,0} \mathbf{v}_{i-r_{j,i}} , \tag{3.82}$$

where $\mathbf{P}_i$ is given by (3.73) and

$$\mathbf{R}_{j,i,0} = \int_0^h e^{\mathbf{A}_i(h-s)} \, ds \, \mathbf{B}_{j,i} . \tag{3.83}$$

If $\mathbf{A}_i^{-1}$ exists, then integration gives

$$\mathbf{R}_{j,i,0} = \left(e^{\mathbf{A}_i h} - \mathbf{I}\right) \mathbf{A}_i^{-1} \mathbf{B}_{j,i} , \tag{3.84}$$

where $\mathbf{I}$ is the $n \times n$ identity matrix. In this case, the coefficient matrix $\mathbf{G}_i$ in (3.75) reads

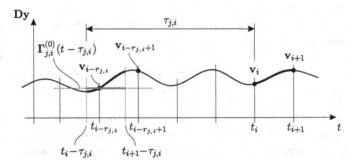

Fig. 3.11 Zeroth-order approximation of the delayed term $\mathbf{D}\mathbf{y}(t - \tau_{j,i})$ by $\boldsymbol{\Gamma}_{j,i}^{(0)}(t - \tau_{j,i}) \equiv \mathbf{v}_{i-r_{j,i}}$, shown by the dashed line.

$$\mathbf{G}_i = \left(\begin{array}{c|cccc} \mathbf{P}_i & 0 & \cdots & 0 & 0 \\ \hline \mathbf{D} & 0 & \cdots & 0 & 0 \\ 0 & \mathbf{I} & \cdots & 0 & 0 \\ \vdots & & \ddots & & \vdots \\ 0 & 0 & \cdots & \mathbf{I} & 0 \end{array} \right) + \sum_{j=1}^{g} \left(\begin{array}{c|ccccccc} \overset{1}{0} & 0 & \cdots & 0 & \overset{r_{j,i}}{\mathbf{R}_{j,i,0}} & 0 & \cdots & \overset{r}{0} \\ \hline 0 & 0 & \cdots & 0 & 0 & 0 & \cdots & 0 \\ 0 & 0 & \cdots & 0 & 0 & 0 & \cdots & 0 \\ \vdots & \vdots & & \vdots & \vdots & \vdots & & \vdots \\ 0 & 0 & \cdots & 0 & 0 & 0 & \cdots & 0 \end{array} \right). \tag{3.85}$$

Here, matrix $\mathbf{R}_{j,i,0}$ is located at the $r_{j,i}$th place within the right upper block of $\mathbf{G}_i$. The approximate monodromy matrix is given by (3.79).

Improved Zeroth-Order Semi-Discretization

An improved version of the zeroth-order method was suggested by Elbeyli and Sun [73] and by Insperger and Stepan [126]. In this case, the approximate system reads

$$\dot{\mathbf{y}}(t) = \mathbf{A}_i \mathbf{y}(t) + \sum_{j=1}^{g} \mathbf{B}_{j,i} \left(\beta_{j,i,0} \mathbf{v}(t_{i-r_{j,i}}) + \beta_{j,i,1} \mathbf{v}(t_{i-r_{j,i}+1}) \right), \quad t \in [t_i, t_{i+1}), \tag{3.86}$$

$$\mathbf{v}(t_i) = \mathbf{D} \mathbf{y}(t_i), \tag{3.87}$$

where

$$\beta_{j,i,0} = \frac{\tau_{j,i} + \frac{1}{2}h - r_{j,i}h}{h}, \qquad \beta_{j,i,1} = \frac{r_{j,i}h + \frac{1}{2}h - \tau_{j,i}}{h}, \tag{3.88}$$

and $r_{j,i} = \text{int}(\tau_{j,i}/h + 1/2)$. Use again the notation $\mathbf{y}_i := \mathbf{y}(t_i)$ and $\mathbf{v}_i := \mathbf{v}(t_i)$, $i \in \mathbb{Z}$. In this case, the delayed term is determined by a weighted average of $\mathbf{v}_{i-r_{j,i}}$ and $\mathbf{v}_{i-r_{j,i}+1}$ in the form

$$\mathbf{\Gamma}_{j,i}^{(0*)}(t - \tau_{j,i}) \equiv \beta_{j,i,0} \mathbf{v}_{i-r_{j,i}} + \beta_{j,i,1} \mathbf{v}_{i-r_{j,i}+1}, \tag{3.89}$$

as shown also in Figure 3.12. Note that $\beta_{j,i,0} + \beta_{j,i,1} = 1$.

The solution over one discrete step can be formulated as

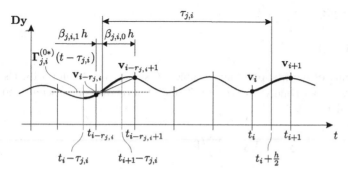

Fig. 3.12 Improved zeroth-order approximation of the delayed term $\mathbf{D}\mathbf{y}(t - \tau_{j,i})$ by $\mathbf{\Gamma}_{j,i}^{(0*)}(t - \tau_{j,i})$, shown by the dashed line.

$$\mathbf{y}_{i+1} = \mathbf{P}_i \mathbf{y}_i + \sum_{j=1}^{g} \mathbf{R}_{j,i,0} \left(\beta_{j,i,0} \mathbf{v}_{i-r_{j,i}} + \beta_{j,i,1} \mathbf{v}_{i-r_{j,i}+1} \right) , \qquad (3.90)$$

where $\mathbf{P}_i$ and $\mathbf{R}_{j,i,0}$ are given by (3.73) and (3.83), respectively. In this case, the coefficient matrix $\mathbf{G}_i$ in (3.75) can be constructed as

$$\mathbf{G}_i = \begin{pmatrix} \mathbf{P}_i & 0 & \cdots & 0 & 0 \\ \hline \mathbf{D} & 0 & \cdots & 0 & 0 \\ 0 & \mathbf{I} & \cdots & 0 & 0 \\ \vdots & & \ddots & & \vdots \\ 0 & 0 & \cdots & \mathbf{I} & 0 \end{pmatrix} + \sum_{j=1}^{g} \overset{\begin{matrix} 1 & \quad r_{j,i}-1 & \ r_{j,i} & \qquad r \end{matrix}}{\begin{pmatrix} 0 & 0 & \cdots & 0 & \beta_{j,i,1}\mathbf{R}_{j,i,0} & \beta_{j,i,0}\mathbf{R}_{j,i,0} & 0 & \cdots & 0 \\ \hline 0 & 0 & \cdots & 0 & 0 & 0 & 0 & \cdots & 0 \\ 0 & 0 & \cdots & 0 & 0 & 0 & 0 & \cdots & 0 \\ \vdots & \vdots & & \vdots & \vdots & \vdots & \vdots & & \vdots \\ 0 & 0 & \cdots & 0 & 0 & 0 & 0 & \cdots & 0 \end{pmatrix}} .$$

$$(3.91)$$

Here, matrices $\beta_{j,i,1}\mathbf{R}_{j,i,0}$ and $\beta_{j,i,0}\mathbf{R}_{j,i,0}$ are located at the $(r_{j,i}-1)$th and $r_{j,i}$th places within the right upper block of $\mathbf{G}_i$, respectively. Again, the approximate monodromy matrix is given by (3.79).

First-Order Semi-Discretization

For the first-order case ($q = 1$), (3.66) and (3.71) give

$$\Gamma_{j,i}^{(1)}(t - \tau_{j,i}) = \beta_{j,i,0}(t)\mathbf{v}(t_{i-r_{j,i}}) + \beta_{j,i,1}(t)\mathbf{v}(t_{i-r_{j,i}+1}) \qquad (3.92)$$

with

$$\beta_{j,i,0}(t) = \frac{\tau_{j,i} + (i - r_{j,i} + 1)h - t}{h} , \qquad \beta_{j,i,1}(t) = \frac{t - (i - r_{j,i})h - \tau_{j,i}}{h} , \qquad (3.93)$$

and $r_{j,i} = \text{int}(\tau_{j,i}/h + 1/2)$. In this case, the approximate system reads

$$\dot{\mathbf{y}}(t) = \mathbf{A}_i \mathbf{y}(t) + \sum_{j=1}^{g} \mathbf{B}_{j,i} \left(\beta_{j,i,0}(t)\mathbf{v}(t_{i-r_{j,i}}) + \beta_{j,i,1}(t)\mathbf{v}(t_{i-r_{j,i}+1}) \right) , \quad t \in [t_i, t_{i+1}) , \quad (3.94)$$

$$\mathbf{v}(t_i) = \mathbf{D}\mathbf{y}(t_i) . \qquad (3.95)$$

The scheme of the first-order approximation is shown in Figure 3.13. Using the notation $\mathbf{y}_i := \mathbf{y}(t_i)$ and $\mathbf{v}_i := \mathbf{v}(t_i)$, $i \in \mathbb{Z}$, the solution over one discrete step can be formulated as

$$\mathbf{y}_{i+1} = \mathbf{P}_i \mathbf{y}_i + \sum_{j=1}^{g} \left(\mathbf{R}_{j,i,0} \mathbf{v}_{i-r_{j,i}} + \mathbf{R}_{j,i,1} \mathbf{v}_{i-r_{j,i}+1} \right) , \qquad (3.96)$$

where $\mathbf{P}_i$ is given by (3.73) and

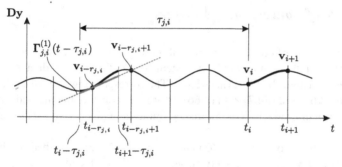

Fig. 3.13 First-order approximation of the delayed term $\mathbf{Dy}(t - \tau_{j,i})$ by $\mathbf{\Gamma}_{j,i}^{(1)}(t - \tau_{j,i})$, shown by the dashed line.

$$\mathbf{R}_{j,i,0} = \int_0^h \frac{\tau_{j,i} - (r_{j,i} - 1)h - s}{h} \, e^{\mathbf{A}_i(h-s)} \, ds \, \mathbf{B}_{j,i} \,, \tag{3.97}$$

$$\mathbf{R}_{j,i,1} = \int_0^h \frac{s - \tau_{j,i} + r_{j,i}h}{h} \, e^{\mathbf{A}_i(h-s)} \, ds \, \mathbf{B}_{j,i} \,. \tag{3.98}$$

If $\mathbf{A}_i^{-1}$ exists, then integration gives

$$\mathbf{R}_{j,i,0} = \left(\mathbf{A}_i^{-1} + \frac{1}{h} \left(\mathbf{A}_i^{-2} - (\tau_{j,i} - (r_{j,i} - 1)h) \mathbf{A}_i^{-1} \right) \left(\mathbf{I} - e^{\mathbf{A}_i h} \right) \right) \mathbf{B}_{j,i}, \tag{3.99}$$

$$\mathbf{R}_{j,i,1} = \left(-\mathbf{A}_i^{-1} + \frac{1}{h} \left(-\mathbf{A}_i^{-2} + (\tau_{j,i} - r_{j,i}h) \mathbf{A}_i^{-1} \right) \left(\mathbf{I} - e^{\mathbf{A}_i h} \right) \right) \mathbf{B}_{j,i}. \tag{3.100}$$

In this case, the coefficient matrix $\mathbf{G}_i$ in (3.75) reads

$$\mathbf{G}_i = \left(\begin{array}{c|cccc} \mathbf{P}_i & \mathbf{0} & \cdots & \mathbf{0} & \mathbf{0} \\ \hline \mathbf{D} & \mathbf{0} & \cdots & \mathbf{0} & \mathbf{0} \\ \mathbf{0} & \mathbf{I} & \cdots & \mathbf{0} & \mathbf{0} \\ \vdots & & \ddots & & \vdots \\ \mathbf{0} & \mathbf{0} & \cdots & \mathbf{I} & \mathbf{0} \end{array} \right) + \sum_{j=1}^{g} \overset{\begin{array}{ccccccccc} \ & \ & 1 & \ & \ & r_{j,i}-1 & r_{j,i} & \ & \ & \ & r \end{array}}{\left(\begin{array}{c|ccccccccc} \mathbf{0} & \mathbf{0} & \cdots & \mathbf{0} & \mathbf{R}_{j,i,1} & \mathbf{R}_{j,i,0} & \mathbf{0} & \cdots & \mathbf{0} \\ \hline \mathbf{0} & \mathbf{0} & \cdots & \mathbf{0} & \mathbf{0} & \mathbf{0} & \mathbf{0} & \cdots & \mathbf{0} \\ \mathbf{0} & \mathbf{0} & \cdots & \mathbf{0} & \mathbf{0} & \mathbf{0} & \mathbf{0} & \cdots & \mathbf{0} \\ \vdots & \vdots & & \vdots & \vdots & \vdots & \vdots & & \vdots \\ \mathbf{0} & \mathbf{0} & \cdots & \mathbf{0} & \mathbf{0} & \mathbf{0} & \mathbf{0} & \cdots & \mathbf{0} \end{array} \right)}. \tag{3.101}$$

Here, matrices $\mathbf{R}_{j,i,1}$ and $\mathbf{R}_{j,i,0}$ are located at the $(r_{j,i} - 1)$th and $r_{j,i}$th places within the right upper block of $\mathbf{G}_i$, respectively. The approximate monodromy matrix is given by (3.79).

3.3.2 Rate of Convergence Estimates

In this section, rate of convergence estimates are given for the zeroth- and the first-order semi-discretization methods. A system with multiple point delays is considered in the form given by (3.62) and (3.63). The corresponding semi-discrete system is given by (3.65) and (3.67) with (3.66). In order to compare the methods of different order, local discretization errors are determined as a function of the discretization step h.

The state spaces of (3.62)–(3.63) and (3.65)–(3.67) are described by the functions $\mathbf{x}_t(\vartheta)$ and $\mathbf{y}_t(\vartheta)$, respectively, where $\vartheta \in [-\sigma, 0]$ with $\sigma = \max(\tau_j(s))$, $j = 1, 2, \ldots, g$, $s \in [0, T]$. The local discretization error over the discretization interval $[0, h]$ is defined as

$$E_{\text{local}} = \|\mathbf{x}_t(h) - \mathbf{y}_t(h)\| , \tag{3.102}$$

where it is assumed that the initial conditions satisfy $\mathbf{x}_t(0) \equiv \mathbf{y}_t(0)$. Here, $\|\cdot\|$ denotes the supremum norm, i.e.,

$$E_{\text{local}} = \sup_{t \in [-\sigma, 0]} |\mathbf{x}_t(h) - \mathbf{y}_t(h)| , \tag{3.103}$$

where $|\cdot|$ denotes the Euclidean norm on $\mathbb{R}^n$. Assume that $h < \sigma$ (note that this is usually the case in practical problems). Utilizing that $\mathbf{x}_t(\vartheta) = \mathbf{x}(t+\vartheta)$, $\mathbf{y}_t(\vartheta) = \mathbf{y}(t+\vartheta)$ and $\mathbf{x}_t(0) \equiv \mathbf{y}_t(0)$, one obtains

$$E_{\text{local}} = \sup_{t \in (0, h]} |\mathbf{x}(t) - \mathbf{y}(t)| . \tag{3.104}$$

For $0 < t \leq h$, the difference $\mathbf{x}(t) - \mathbf{y}(t)$ can be expanded as

$$\mathbf{x}(t) - \mathbf{y}(t) = \int_0^t \mathbf{A}(s)\mathbf{x}(s) - \mathbf{A}_0 \mathbf{y}(s) \, ds \tag{3.105}$$

$$+ \sum_{j=1}^g \int_0^t \mathbf{B}_j(s)\mathbf{u}(s - \tau_j(s)) - \mathbf{B}_{j,0} \mathbf{\Gamma}_{j,0}^{(q)}(s - \tau_{j,0}) \, ds$$

$$= \underbrace{\int_0^t \mathbf{A}(s)\mathbf{x}(s) - \mathbf{A}_0 \mathbf{x}(s) \, ds + \int_0^t \mathbf{A}_0 \mathbf{x}(s) - \mathbf{A}_0 \mathbf{y}(s) \, ds}_{=: \, \mathbf{J}_0(t)}$$

$$+ \sum_{j=1}^g \Bigg(\underbrace{\int_0^t \mathbf{B}_j(s)\mathbf{u}(s - \tau_j(s)) - \mathbf{B}_{j,0}\mathbf{u}(s - \tau_j(s)) \, ds}_{=: \, \mathbf{J}_{j,1}(t)}$$

$$+ \underbrace{\int_0^t \mathbf{B}_{j,0}\mathbf{u}(s - \tau_j(s)) - \mathbf{B}_{j,0}\mathbf{u}(s - \tau_{j,0}) \, ds}_{=: \, \mathbf{J}_{j,2}(t)}$$

$$+ \underbrace{\int_0^t \mathbf{B}_{j,0} \mathbf{D} \mathbf{x}(s - \tau_{j,0}) - \mathbf{B}_{j,0} \mathbf{D} \mathbf{y}(s - \tau_{j,0}) \, ds}_{=: \mathbf{J}_{j,3}(t)}$$

$$+ \underbrace{\int_0^t \mathbf{B}_{j,0} \mathbf{D} \mathbf{y}(s - \tau_{j,0}) - \mathbf{B}_{j,0} \mathbf{\Gamma}_{j,0}^{(q)}(s - \tau_{j,i}) \, ds}_{=: \mathbf{J}_{j,4}(t)} \Bigg).$$

A measure for the difference $\mathbf{x}(t) - \mathbf{y}(t)$ can be given as

$$E(t) = \|\mathbf{x}(t) - \mathbf{y}(t)\| \le \left\| \mathbf{J}_0(t) + \sum_{j=1}^g \sum_{k=1}^4 \mathbf{J}_{j,k}(t) \right\| + \int_0^t K \|\mathbf{x}(s) - \mathbf{y}(s)\| \, ds , \quad (3.106)$$

where $K = \|\mathbf{A}_0\|$. Due to the Gronwall inequality, (3.106) yields

$$E(t) = \|\mathbf{x}(t) - \mathbf{y}(t)\| \le \left\| \mathbf{J}_0(t) + \sum_{j=1}^g \sum_{k=1}^4 \mathbf{J}_{j,k}(t) \right\| e^{Kt}. \quad (3.107)$$

If $h < \tau_{j,i}$, then, due to the initial assumption $\mathbf{x}_t(0) \equiv \mathbf{y}_t(0)$, the term $\mathbf{J}_{j,3}(t)$ is equal to 0. For the analysis of the other terms $\mathbf{J}_0(t), \mathbf{J}_{j,k}(t), j = 1, 2, \ldots, g, k = 1, 2, 4$, consider the Taylor expansions

$$\mathbf{x}(t) = \tilde{\mathbf{x}}_0 + \tilde{\mathbf{x}}_1 t + \tilde{\mathbf{x}}_2 t^2 + \cdots , \quad (3.108)$$

$$\mathbf{y}(t) = \tilde{\mathbf{y}}_0 + \tilde{\mathbf{y}}_1 t + \tilde{\mathbf{x}}_2 t^2 + \cdots , \quad (3.109)$$

$$\mathbf{A}(t) = \tilde{\mathbf{A}}_0 + \tilde{\mathbf{A}}_1 t + \tilde{\mathbf{A}}_2 t^2 + \cdots , \quad (3.110)$$

$$\mathbf{B}(t) = \tilde{\mathbf{B}}_0 + \tilde{\mathbf{B}}_1 t + \tilde{\mathbf{B}}_2 t^2 + \cdots , \quad (3.111)$$

$$\tau_j(t) = \tilde{\tau}_{j,0} + \tilde{\tau}_{j,1} t + \tilde{\tau}_{j,2} t^2 + \cdots . \quad (3.112)$$

Note that the initial assumption $\mathbf{x}_t(0) \equiv \mathbf{y}_t(0)$ implies $\tilde{\mathbf{x}}_0 = \tilde{\mathbf{y}}_0$.

According to (3.71), $|\tau_{j,0} - r_{j,0} h| \le q/2$, where q is the order of the approximation of the delayed term. Using (3.70), $\tau_{j,0}$ can be expanded as

$$\tau_{j,0} = \tilde{\tau}_{j,0} + \frac{1}{2} \tilde{\tau}_{j,1} h + \tfrac{1}{3} \tilde{\tau}_{j,2} h^2 + \cdots . \quad (3.113)$$

These imply that

$$|\tilde{\tau}_{j,0} - r_{j,0} h| \le q/2 + O(h) . \quad (3.114)$$

Using the above Taylor expansions and taking into account (3.63) and (3.67), applying (3.114), and noting that $t \le h$, the magnitude of the terms $\mathbf{J}_0(h)$ and $\mathbf{J}_{j,k}(h)$, $j = 1, 2, \ldots, g, k = 1, 2, 4$, can be estimated with respect to the discretization step h.

A long but straightforward calculation gives that $\mathbf{J}_0(h) = O(h^3)$, and it does not depend on q. The other terms, however, do depend on q, such that if　$q = 0$　then　$\mathbf{J}_{j,1}(h) = O(h^3)$,　$\mathbf{J}_{j,2}(h) = O(h^3)$,　$\mathbf{J}_{j,4}(h) = O(h^2)$;

if　$q = 1$　then　$\mathbf{J}_{j,1}(h) = O(h^3)$,　$\mathbf{J}_{j,2}(h) = O(h^3)$,　$\mathbf{J}_{j,4}(h) = O(h^3)$.

Using (3.107) and (3.104), the local discretization error can be given as

$$\text{if} \quad q = 0 \quad \text{then} \quad E_{\text{local}} = E(h) = O(h^2),$$

$$\text{if} \quad q = 1 \quad \text{then} \quad E_{\text{local}} = E(h) = O(h^3).$$

If $q > 1$, then the order of the term $\mathbf{J}_{j,4}(h)$ increases, but the terms $\mathbf{J}_0(h)$, $\mathbf{J}_{j,1}(h)$, and $\mathbf{J}_{j,2}(h)$ remain or order 3. Consequently, $E_{\text{local}} = O(h^3)$ for all $q > 1$. In order to achieve higher-order convergence for $q > 1$, the approximation of the time-periodic terms should be improved. For this purpose, a higher-order Magnus expansion should be used for the approximation of the periodic terms $\mathbf{A}(t)$, $\mathbf{B}_j(t)$, and $\tau_j(t)$ instead of the piecewise constant approximations (3.68), (3.69), (3.70) (see, e.g., [175, 138, 139, 46]). Note that approximations (3.68), (3.69), (3.70) correspond to the first-order Magnus expansion. Here, we investigate only the cases $q = 0$ and $q = 1$, for which the piecewise constant approximation is appropriate.

3.4 Numerical Issues

The main drawback of discretization techniques is that the computational cost increases for low discretization steps. The computational efficiency of the semi-discretization method can be increased by applying numerical techniques in the matrix multiplication, in the eigenvalue computation, or in the discretization scheme itself. In this section, some of these techniques are presented to speed up the computations for large matrix dimensions. First, an effective form of matrix multiplication is shown utilizing the sparse structure of the matrices. Then, it is shown that for some simple cases, the monodromy matrix can be obtained in one step without actually performing the matrix multiplications. After that, some comments are given for the eigenvalue computations. Finally, an alternative semi-discretization technique is presented.

3.4.1 Effective Matrix Multiplication

Calculation of the monodromy matrix $\mathbf{\Phi}$ by the semi-discretization method requires $(p - 1)$ multiplications of the $(n + rm) \times (n + rm)$ matrices $\mathbf{G}_i$, $i = 1, 2, \ldots, p$ (see (3.61) or (3.79)). The period resolution p is chosen such that $T = ph$, where T is the principal period of the system, and h is the discretization step. The delay resolution r is defined as $r = \max(r_{j,i})$ with $r_{j,i} = \text{int}(\tau_{j,i}/h + q/2) = \text{int}(p\tau_{j,i}/T + q/2)$ and $\tau_{j,i}$ being defined in (3.70). Consequently, $r = \text{int}(p\tau_{\text{max}}/T + q/2)$, where $\tau_{\text{max}} = \max(\tau_{j,i})$, that is, the delay resolution r is nearly proportional to the period

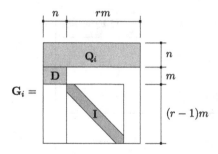

Fig. 3.14 Structure of matrix $\mathbf{G}_i$ in case of distributed delays. The nonzero elements are indicated by gray shading; $\mathbf{I}$ denotes the identity matrix.

resolution p. A single multiplication of two $(n + rm) \times (n + rm)$ matrices requires $N_{1,\text{full}} = 2(n + rm)^3 - (n + rm)^2$ arithmetic operations (i.e., $(n + rm)^3$ floating-point multiplications and $(n + rm - 1)(n + rm)^2$ floating-point additions). The number of arithmetic operations to obtain the monodromy matrix $\mathbf{\Phi}$ by full matrix multiplication is

$$N_{\Phi,\text{full}} = (p - 1)N_{1,\text{full}} = (p - 1)\left(2(n + rm)^3 - (n + rm)^2\right) . \qquad (3.115)$$

Introducing the ratio $\varepsilon = r/p \approx \tau_{\max}/T$ gives

$$N_{\Phi,\text{full}} = (p - 1)\left(2(n + \varepsilon pm)^3 - (n + \varepsilon pm)^2\right) = 2\varepsilon^3 m^3 p^4 + O(p^3) . \qquad (3.116)$$

Thus, the order of the computational effort with respect to the period resolution p is $O(p^4)$.

In order to reduce computational costs, the sparsity of matrices $\mathbf{G}_i$ can be utilized during the matrix multiplications, as shown by Henninger and Eberhard [106]. According to (3.59), transition matrix $\mathbf{G}_i$ can be written in the form

$$\mathbf{G}_i = \begin{pmatrix} \mathbf{P}_i & \tilde{\mathbf{R}}_{1,i} & \cdots & \tilde{\mathbf{R}}_{r-1,i} & \tilde{\mathbf{R}}_{r,i} \\ \mathbf{D} & \mathbf{0} & \cdots & \mathbf{0} & \mathbf{0} \\ \mathbf{0} & \mathbf{I} & \cdots & \mathbf{0} & \mathbf{0} \\ \vdots & & \ddots & & \vdots \\ \mathbf{0} & \mathbf{0} & \cdots & \mathbf{I} & \mathbf{0} \end{pmatrix}, \qquad (3.117)$$

where some of the matrices $\tilde{\mathbf{R}}_{j,i}$, $j = 1, 2, \ldots, r$, might be the zero matrix depending on the delays in the system and on the order of discretization. The structure of this matrix is presented in Figure 3.14, where

$$\mathbf{Q}_i = \begin{pmatrix} \mathbf{P}_i & \tilde{\mathbf{R}}_{1,i} & \cdots & \tilde{\mathbf{R}}_{r-1,i} & \tilde{\mathbf{R}}_{r,i} \end{pmatrix} . \qquad (3.118)$$

If the matrix product $\mathbf{F}_i := \mathbf{G}_{i-1}\mathbf{G}_{i-2} \cdots \mathbf{G}_0$ is given, then $\mathbf{F}_{i+1} = \mathbf{G}_i\mathbf{G}_{i-1} \cdots \mathbf{G}_0$ can be calculated in three steps, as shown in Figure 3.15. The first n rows are obtained by the product $\mathbf{Q}_i\mathbf{F}_i$, i.e.,

$$\mathbf{F}_{i+1}^{\langle 1:n \rangle} = \mathbf{Q}_i\mathbf{F}_i . \qquad (3.119)$$

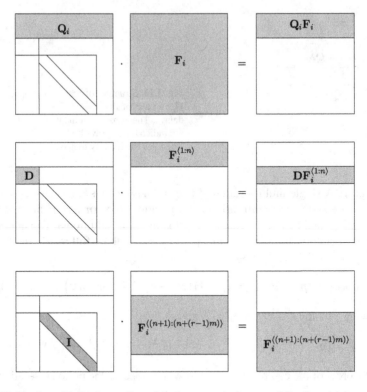

Fig. 3.15 Matrix multiplication performed in three steps in the case of distributed delay.

Here, the notation $\mathbf{M}^{\langle k:l \rangle}$ is used for the submatrix containing the rows from k to l of matrix $\mathbf{M}$. Rows $n + 1, \ldots, n + m$ are given as $\mathbf{DF}_i^{\langle 1:n \rangle}$, i.e.,

$$\mathbf{F}_{i+1}^{\langle (n+1):(n+m) \rangle} = \mathbf{DF}_i^{\langle 1:n \rangle} . \tag{3.120}$$

The remaining part of $\mathbf{F}_{i+1}$ is obtained by a simple shift of the block $\mathbf{F}_i^{\langle (n+1):(n+(r-1)m) \rangle}$, i.e.,

$$\mathbf{F}_{i+1}^{\langle (n+m+1):(n+rm) \rangle} = \mathbf{F}_i^{\langle (n+1):(n+(r-1)m) \rangle} . \tag{3.121}$$

In this way, the multiplication of two $(n + rm) \times (n + rm)$ matrices is reduced to a multiplication of an $n \times (n + rm)$ matrix by an $(n + rm) \times (n + rm)$ one and an $m \times n$ matrix by an $n \times (n + rm)$ one. The number of the corresponding arithmetic operations is

$$N_{1,\text{eff,dd}} = 2(n + rm)^2 n - (n + rm)n + 2nm(n + rm) - m(n + rm) . \tag{3.122}$$

Here, the subscript eff refers to effective matrix multiplication, and dd refers to distributed delay.

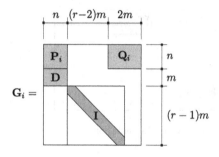

Fig. 3.16 Structure of matrix G_i in the case of a single point delay and treated by the first-order semi-discretization. The nonzero elements are indicated by gray shading; **I** denotes the identity matrix.

The monodromy matrix is obtained after $(p-1)$ repeated matrix multiplications, i.e., $\mathbf{\Phi} = \mathbf{F}_p$. Using the ratio $\varepsilon = r/p$, the number of arithmetic operations to compute $\mathbf{\Phi}$ by the effective matrix multiplication is

$$N_{\Phi,\text{eff,dd}} = (p-1)\big(2(n+\varepsilon pm)^2 n - (n+\varepsilon pm)n + 2nm(n+\varepsilon pm) - m(n+\varepsilon pm)\big)$$
$$= 2nm^2\varepsilon^2 p^3 + O(p^2)\,, \tag{3.123}$$

that is, the order of the computational effort with respect to p is $O(p^3)$.

Note that the above analysis is valid for systems with distributed delays, where submatrix $\mathbf{Q}_i$ is a full matrix. In the case of a system with point delays, submatrix $\mathbf{Q}_i$ is sparse itself, which facilitates further simplifications in the computation process.

Consider a system with a single point delay and employ the first-order semi-discretization method. According to (3.101), transition matrix $\mathbf{G}_i$ can be written in the form

$$\mathbf{G}_i = \begin{pmatrix} \mathbf{P}_i & \mathbf{0} & \cdots & \mathbf{0} & \mathbf{R}_{i,1} & \mathbf{R}_{i,0} \\ \mathbf{D} & \mathbf{0} & \cdots & \mathbf{0} & \mathbf{0} & \mathbf{0} \\ \mathbf{0} & \mathbf{I} & \cdots & \mathbf{0} & \mathbf{0} & \mathbf{0} \\ \vdots & & & & & \vdots \\ \mathbf{0} & \mathbf{0} & \cdots & \mathbf{0} & \mathbf{I} & \mathbf{0} \end{pmatrix}. \tag{3.124}$$

The structure of this matrix is presented in Figure 3.16, where

$$\mathbf{Q}_i = \begin{pmatrix} \mathbf{R}_{i,1} & \mathbf{R}_{i,0} \end{pmatrix}. \tag{3.125}$$

In this case, the first step of the matrix multiplication is reduced to a multiplication of an $n \times n$ matrix by an $n \times (n + rm)$ one and an $n \times 2m$ matrix by a $2m \times (n + rm)$ one, as shown in Figure 3.17. In this case, the first n rows are obtained as

$$\mathbf{F}_{i+1}^{\langle 1:n\rangle} = \mathbf{P}_i\mathbf{F}_i^{\langle 1:n\rangle} + \mathbf{Q}_i\mathbf{F}_i^{\langle (n+(r-2)m+1):(n+rm)\rangle}\,. \tag{3.126}$$

The number of arithmetic operations for a single matrix multiplication is

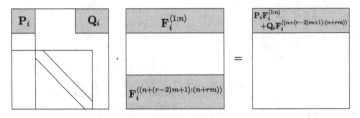

Fig. 3.17 The first step of the matrix multiplication in the case of a single point delay.

$$N_{1,\text{eff,pd}} = 2n^2(n + rm) - n(n + rm) + 8m^2(n + rm) - 2m(n + rm)$$
$$+ n(n + rm) + 2nm(n + rm) - m(n + rm) , \quad (3.127)$$

where the subscript pd refers to point delay. Consequently, the number of arithmetic operations to compute $\mathbf{\Phi}$ is

$$N_{\mathbf{\Phi},\text{eff,pd}} = (p - 1)\big(2n^2(n + \varepsilon pm) - n(n + \varepsilon pm) + 8m^2(n + \varepsilon pm)$$
$$- 2m(n + \varepsilon pm) + n(n + \varepsilon pm) + 2nm(n + \varepsilon pm) - m(n + \varepsilon pm)\big)$$
$$= \varepsilon m \big(2n^2 + 8m^2 - 3m + 2nm\big) p^2 + O(p) , \quad (3.128)$$

where $\varepsilon = r/p$. Thus, the order of the computational effort with respect to the period resolution p is $O(p^2)$.

Note that the same or similar computational shortcuts can be attained by employing sparse matrix multiplication routines that avoid the multiplication of zero entries. In general, these routines can be used to handle matrices of any (unknown) sparsity, which requires some extra computational effort. In the current particular problem, however, the sparse structures of matrices $\mathbf{G}_i$ are already known. Thus, the costs for determining their sparsity can be saved. The effective matrix multiplication therefore provides a slightly faster computation than other sparse matrix multiplications.

3.4.2 Construction of the Monodromy Matrix in One Step

Due to the special structure of matrices $\mathbf{G}_i$, the $(p - 1)$ matrix multiplications can be performed in one step, which can result in further computational cost savings as suggested by Zatarain and Dombovari [302, 67]. To demonstrate this feature, a system with a single point delay is considered with the first-order semi-discretization method for the special case $r = p$. According to (3.101), transition matrix $\mathbf{G}_i$ can be written in the form

$$
\mathbf{G}_i = \begin{pmatrix}
\mathbf{P}_i & \mathbf{0} & \cdots & \mathbf{0} & \mathbf{R}_{i,1} & \mathbf{R}_{i,0} \\
\mathbf{D} & \mathbf{0} & \cdots & \mathbf{0} & \mathbf{0} & \mathbf{0} \\
\mathbf{0} & \mathbf{I} & \cdots & \mathbf{0} & \mathbf{0} & \mathbf{0} \\
\vdots & & & & & \vdots \\
\mathbf{0} & \mathbf{0} & \cdots & \mathbf{0} & \mathbf{I} & \mathbf{0}
\end{pmatrix} .
\tag{3.129}
$$

Successive matrix multiplication gives

$$
\mathbf{F}_2 = \mathbf{G}_1 \mathbf{G}_0 = \begin{pmatrix}
\mathbf{P}_1 \mathbf{P}_0 & \mathbf{0} & \mathbf{0} & \cdots & \mathbf{0} & \mathbf{R}_{1,1} & \mathbf{P}_1 \mathbf{R}_{0,1} + \mathbf{R}_{1,0} & \mathbf{P}_1 \mathbf{R}_{0,0} \\
\mathbf{D}\mathbf{P}_0 & \mathbf{0} & \mathbf{0} & \cdots & \mathbf{0} & \mathbf{0} & \mathbf{D}\mathbf{R}_{0,1} & \mathbf{D}\mathbf{R}_{0,0} \\
\mathbf{D} & \mathbf{0} & \mathbf{0} & \cdots & \mathbf{0} & \mathbf{0} & \mathbf{0} & \mathbf{0} \\
\mathbf{0} & \mathbf{I} & \mathbf{0} & \cdots & \mathbf{0} & \mathbf{0} & \mathbf{0} & \mathbf{0} \\
\vdots & & & & & & & \vdots \\
\mathbf{0} & \mathbf{0} & \mathbf{0} & \cdots & \mathbf{0} & \mathbf{I} & \mathbf{0} & \mathbf{0}
\end{pmatrix} ,
\tag{3.130}
$$

$$
\mathbf{F}_3 = \begin{pmatrix}
\mathbf{P}_2\mathbf{P}_1\mathbf{P}_0 & \mathbf{0} & \mathbf{0} & \cdots & \mathbf{R}_{2,1} & \mathbf{P}_2\mathbf{R}_{1,1} + \mathbf{R}_{2,0} & \mathbf{P}_2(\mathbf{P}_1\mathbf{R}_{0,1} + \mathbf{R}_{1,0}) & \mathbf{P}_2\mathbf{P}_1\mathbf{R}_{0,0} \\
\mathbf{D}\mathbf{P}_1\mathbf{P}_0 & \mathbf{0} & \mathbf{0} & \cdots & \mathbf{0} & \mathbf{D}\mathbf{R}_{1,1} & \mathbf{D}(\mathbf{P}_1\mathbf{R}_{0,1} + \mathbf{R}_{1,0}) & \mathbf{D}\mathbf{P}_1\mathbf{R}_{0,0} \\
\mathbf{D}\mathbf{P}_0 & \mathbf{0} & \mathbf{0} & \cdots & \mathbf{0} & \mathbf{0} & \mathbf{R}_{0,1} & \mathbf{D}\mathbf{R}_{0,0} \\
\mathbf{D} & \mathbf{0} & \mathbf{0} & \cdots & \mathbf{0} & \mathbf{0} & \mathbf{0} & \mathbf{0} \\
\mathbf{0} & \mathbf{I} & \mathbf{0} & \cdots & \mathbf{0} & \mathbf{0} & \mathbf{0} & \mathbf{0} \\
\vdots & & & & & & & \vdots \\
\mathbf{0} & \mathbf{0} & \mathbf{0} & \cdots & \mathbf{I} & \mathbf{0} & \mathbf{0} & \mathbf{0}
\end{pmatrix} .
\tag{3.131}
$$

After $(r-1)$ multiplications, the monodromy matrix reads

$$
\boldsymbol{\Phi} = \mathbf{F}_r = \begin{pmatrix}
\mathbf{P}_{r-1}\dots\mathbf{P}_0 + \mathbf{R}_{r-1,1}\mathbf{D} & \mathbf{P}_{r-1}\mathbf{R}_{r-2,1} + \mathbf{R}_{r-1,0} & \mathbf{P}_{r-1}(\mathbf{P}_{r-2}\mathbf{R}_{r-3,1} + \mathbf{R}_{r-2,0}) & \cdots \\
\mathbf{D}\mathbf{P}_{r-2}\dots\mathbf{P}_0 & \mathbf{D}\mathbf{R}_{r-2,1} & \mathbf{D}(\mathbf{P}_{r-2}\mathbf{R}_{r-3,1} + \mathbf{R}_{r-2,0}) & \cdots \\
\vdots & & & \\
\mathbf{D}\mathbf{P}_3\mathbf{P}_2\mathbf{P}_1\mathbf{P}_0 & \mathbf{0} & \mathbf{0} & \cdots \\
\mathbf{D}\mathbf{P}_2\mathbf{P}_1\mathbf{P}_0 & \mathbf{0} & \mathbf{0} & \cdots \\
\mathbf{D}\mathbf{P}_1\mathbf{P}_0 & \mathbf{0} & \mathbf{0} & \cdots \\
\mathbf{D}\mathbf{P}_0 & \mathbf{0} & \mathbf{0} & \cdots \\
\mathbf{D} & \mathbf{0} & \mathbf{0} & \cdots
\end{pmatrix}
$$

$$
\begin{pmatrix}
\cdots & \mathbf{P}_{r-1}\dots\mathbf{P}_3(\mathbf{P}_2\mathbf{R}_{1,1} + \mathbf{R}_{2,0}) & \mathbf{P}_{r-1}\dots\mathbf{P}_2(\mathbf{P}_1\mathbf{R}_{0,1} + \mathbf{R}_{1,0}) & \mathbf{P}_{r-1}\dots\mathbf{P}_1\mathbf{R}_{0,0} \\
\cdots & \mathbf{D}\mathbf{P}_{r-2}\dots\mathbf{P}_3(\mathbf{P}_2\mathbf{R}_{1,1} + \mathbf{R}_{2,0}) & \mathbf{D}\mathbf{P}_{r-2}\dots\mathbf{P}_2(\mathbf{P}_1\mathbf{R}_{0,1} + \mathbf{R}_{1,0}) & \mathbf{D}\mathbf{P}_{r-2}\dots\mathbf{P}_1\mathbf{R}_{0,0} \\
\vdots & & & \vdots \\
\cdots & \mathbf{D}\mathbf{P}_3(\mathbf{P}_2\mathbf{R}_{1,1} + \mathbf{R}_{2,0}) & \mathbf{D}\mathbf{P}_3\mathbf{P}_2(\mathbf{P}_1\mathbf{R}_{0,1} + \mathbf{R}_{1,0}) & \mathbf{D}\mathbf{P}_3\mathbf{P}_2\mathbf{P}_1\mathbf{R}_{0,0} \\
\cdots & \mathbf{D}(\mathbf{P}_2\mathbf{R}_{1,1} + \mathbf{R}_{2,0}) & \mathbf{D}\mathbf{P}_2(\mathbf{P}_1\mathbf{R}_{0,1} + \mathbf{R}_{1,0}) & \mathbf{D}\mathbf{P}_2\mathbf{P}_1\mathbf{R}_{0,0} \\
\cdots & \mathbf{D}\mathbf{R}_{1,1} & \mathbf{D}(\mathbf{P}_1\mathbf{R}_{0,1} + \mathbf{R}_{1,0}) & \mathbf{D}\mathbf{P}_1\mathbf{R}_{0,0} \\
\cdots & \mathbf{0} & \mathbf{D}\mathbf{R}_{0,1} & \mathbf{D}\mathbf{R}_{0,0} \\
\cdots & \mathbf{0} & \mathbf{0} & \mathbf{0}
\end{pmatrix} .
\tag{3.132}
$$

Table 3.1 Number of arithmetic operations to determine the monodromy matrix by different matrix multiplication techniques for a system with a single point delay treated by the first-order semi-discretization method such that $r = p$.

	method	$m = 1$	$m = 2$	$m = 3$
$n = 1$	full multiplication	$2p^4 + O(p^3)$	$16p^4 + O(p^3)$	$54p^4 + O(p^3)$
	effective multiplication	$9p^2 + O(p)$	$64p^2 + O(p)$	$213p^2 + O(p)$
	Φ in one step	$p^2 + O(p)$	$3p^2 + O(p)$	$6p^2 + O(p)$
$n = 2$	full multiplication	$2p^4 + O(p^3)$	$16p^4 + O(p^3)$	$54p^4 + O(p^3)$
	effective multiplication	$17p^2 + O(p)$	$84p^2 + O(p)$	$249p^2 + O(p)$
	Φ in one step	$\frac{9}{2}p^2 + O(p)$	$12p^2 + O(p)$	$\frac{45}{2}p^2 + O(p)$
$n = 3$	full multiplication	$2p^4 + O(p^3)$	$16p^4 + O(p^3)$	$54p^4 + O(p^3)$
	effective multiplication	$29p^2 + O(p)$	$112p^2 + O(p)$	$297p^2 + O(p)$
	Φ in one step	$10p^2 + O(p)$	$25p^2 + O(p)$	$45p^2 + O(p)$

This matrix can be constructed using the matrices $\mathbf{D}$, $\mathbf{P}_i$, $\mathbf{R}_{i,0}$, and $\mathbf{R}_{i,1}$, $i = 0, 1, \ldots, r - 1$, starting from the bottom right corner. First, the matrix $\mathbf{P}_1 \mathbf{R}_{0,0}$ can be computed, then $\mathbf{P}_2 \mathbf{P}_1 \mathbf{R}_{0,0}$, $\mathbf{P}_3 \mathbf{P}_2 \mathbf{P}_1 \mathbf{R}_{0,0}$, etc. These matrices should then be multiplied by $\mathbf{D}$ (from the left) to obtain the elements in the last column except the one in the first line. The column before the last one can be determined step by step by computing $(\mathbf{P}_1 \mathbf{R}_{0,1} + \mathbf{R}_{1,0})$, $\mathbf{P}_2(\mathbf{P}_1 \mathbf{R}_{0,1} + \mathbf{R}_{1,0})$, $\mathbf{P}_3 \mathbf{P}_2(\mathbf{P}_1 \mathbf{R}_{0,1} + \mathbf{R}_{1,0})$, etc., and finally all (except the one in the first line) multiplied by $\mathbf{D}$. All the other columns can be determined in a similar way. The computation of the monodromy matrix $\mathbf{\Phi} = \mathbf{F}_r$ in this case requires

$$N_{\Phi,1\text{step,pd}} = (p - 1)\left(2n^3 - n^2 + 2n^2 m - nm + 2n^2 m - nm + 2nm^2 - m^2\right)$$

$$+ 2n^2 m - n^2 + n^2 + \sum_{j=1}^{p-1}\left((j - 1)\left(2n^2 m - nm + 2nm^2 - m^2\right) + nm\right)$$

$$= \frac{1}{2}m(2n - 1)(n + m)p^2 + O(p) \tag{3.133}$$

arithmetic operations. Similarly to the effective matrix multiplication (see (3.128)), the order of the computational effort with respect to the period resolution p is $O(p^2)$. However, the coefficient of p^2 in (3.133) is less than in (3.128).

A comparison of the computational costs for the full matrix multiplication, for the effective matrix multiplication, and for the determination of the monodromy matrix in one step is shown in Table 3.1 for different system dimensions n and m. For this comparison, a system with a single point delay being equal to the principal period (i.e., $\tau = T$) is considered, which is treated by the first-order semi-discretization method such that the delay resolution is equal to the period resolution (i.e., $r = p$).

The above construction of the monodromy matrix is valid only for the particular case $r = p$. If $r > p$, then the method can be used in the same manner. In this case, the matrix contains fewer nonzero entries, and the rows from $(n + pm + 1)$ to $n + rm$ contain only zeros and ones, similarly to (3.131). If $r < p$, then the monodromy matrix can be determined by multiplications in groups of r matrices such that

$$\mathbf{H}_{r-1,0} = (\mathbf{G}_{r-1} \ldots \mathbf{G}_1 \mathbf{G}_0) \,, \tag{3.134}$$

$$\mathbf{H}_{2r-1,r} = (\mathbf{G}_{2r-1} \ldots \mathbf{G}_{r+1} \mathbf{G}_r) \,, \tag{3.135}$$

$$\vdots \tag{3.136}$$

$$\mathbf{H}_{p-1,kr} = (\mathbf{G}_{p-1} \ldots \mathbf{G}_{kr+1} \mathbf{G}_{kr}) \,, \tag{3.137}$$

where $k = \text{int}(p/r)$. In this case, the monodromy matrix is obtained by k multiplications of the above $(n + rm) \times (n + rm)$ matrices as

$$\mathbf{\Phi} = \mathbf{H}_{p-1,kr} \, \mathbf{H}_{kr-1,(k-1)r} \, \cdots \, \mathbf{H}_{r-1,0} \,. \tag{3.138}$$

This additional multiplication requires, however, some extra computational effort compared to the case $r \geq p$.

3.4.3 Eigenvalue Computation

In addition to the multiplication of large matrices, the calculation of the eigenvalues of the monodromy matrix $\mathbf{\Phi}$ is another operation that requires nonnegligible computational resources. There are several numerical algorithms to determine the largest (in modulus) eigenvalue of a matrix; see, e.g., the Arnoldi iteration, the Lanczos method, and their different implementations [11, 167, 8, 59, 48, 168].

In the Matlab environment, the eigenvalues of an $n \times n$ matrix $\mathbf{A}$ can be computed directly either by the eig or by the eigs command. The command eig uses LA-PACK routines to compute eigenvalues and eigenvectors [8], while the command eigs uses ARPACK routines [168]. A comparison of the computational time for both commands is shown in Figure 3.18 for matrices of different sizes. If the size of the matrix is less than 60, then eig is faster than eigs. However, for matrices of size larger than 60, eigs is superior to eig. This effect should also be considered in the computational codes.

3.4.4 An Alternative Semi-Discretization Method

There are other similar numerical techniques to determine the stability of time-periodic DDEs. Here, an alternative semi-discretization method introduced by Ding et al. [65, 66] is presented. Consider the DDE of the form

$$\dot{\mathbf{x}}(t) = \mathbf{A}\mathbf{x}(t) + \mathbf{B}(t)\mathbf{x}(t) + \mathbf{B}(t)\mathbf{x}(t - \tau) \,, \tag{3.139}$$

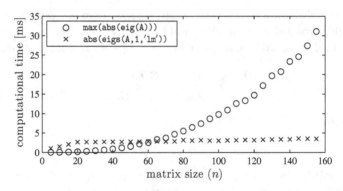

Fig. 3.18 Computational times for determining the maximum (in modulus) eigenvalue of a random $n \times n$ matrix using Matlab's `eig` and `eigs` commands.

where $\mathbf{x}(t) \in \mathbb{R}^n$, $\mathbf{A}$ is an $n \times n$ constant matrix, and $\mathbf{B}(t + \tau) = \mathbf{B}(t)$ is an $n \times n$ time-periodic matrix. This equation describes the dynamics of milling processes with zero helix angle and uniform tooth pitches. Therefore, it contains a single constant time-delay that is just equal to the principal period of the system. According to the method of Ding et al. [65, 66], the discretized equation reads

$$\dot{\mathbf{y}}(t) = \mathbf{A}\mathbf{y}(t) + \tilde{\mathbf{B}}(t)\tilde{\mathbf{y}}(t) + \tilde{\mathbf{B}}(t)\tilde{\mathbf{y}}(t - \tau) , \qquad t \in [t_i, t_{i+1}) , \tag{3.140}$$

where $t_i = ih$, with h being the discretization step and

$$\tilde{\mathbf{B}}(t) = \mathbf{B}_i + \frac{\mathbf{B}_{i+1} - \mathbf{B}_i}{h}(t - t_i) , \tag{3.141}$$

$$\tilde{\mathbf{y}}(t) = \mathbf{y}_i + \frac{\mathbf{y}_{i+1} - \mathbf{y}_i}{h}(t - t_i) , \tag{3.142}$$

$$\tilde{\mathbf{y}}(t - \tau) = \mathbf{y}_{i-p} + \frac{\mathbf{y}_{i-p+1} - \mathbf{y}_{i-p}}{h}(t - t_i) , \tag{3.143}$$

are the linear Lagrange polynomial approximations of $\mathbf{B}(t)$, $\mathbf{y}(t)$, and $\mathbf{y}(t-\tau)$, respectively, with $\mathbf{B}_i = \mathbf{B}(t_i)$ and $\mathbf{y}_i = \mathbf{y}(t_i)$. The discretization step is determined such that $h = \tau/p$, where p is the approximation parameter. The key point of this discretization technique is that both the delayed term $\mathbf{B}(t)\mathbf{x}(t - \tau)$ and the nondelayed term $\mathbf{B}(t)\mathbf{x}(t)$ of periodic coefficient are discretized, while the nondelayed term $\mathbf{A}\mathbf{x}(t)$ of constant coefficient is left in its original form, which is associated with the exact exponential solution $e^{\mathbf{A}(t-t_i)}\mathbf{x}(t_i)$.

Solving (3.140) as an ODE over the discretization period $[t_i, t_{i+1}]$, one ends up with

$$\mathbf{y}_{i+1} = \left(e^{\mathbf{A}h} + \mathbf{F}_i\right)\mathbf{y}_i + \mathbf{F}_{i+1}\mathbf{y}_{i+1} - \mathbf{F}_i\mathbf{y}_{i-p} - \mathbf{F}_{i+1}\mathbf{y}_{i-p+1} , \tag{3.144}$$

where

$$\mathbf{F}_i = \left(\mathbf{M}_0 - \tfrac{2}{h}\mathbf{M}_1 + \tfrac{1}{h^2}\mathbf{M}_2\right)\mathbf{B}_i + \left(\tfrac{1}{h}\mathbf{M}_1 - \tfrac{1}{h^2}\mathbf{M}_2\right)\mathbf{B}_{i+1}, \tag{3.145}$$

$$\mathbf{F}_{i+1} = \left(\tfrac{1}{h}\mathbf{M}_1 - \tfrac{1}{h^2}\mathbf{M}_2\right)\mathbf{B}_i + \left(\tfrac{1}{h^2}\mathbf{M}_2\right)\mathbf{B}_{i+1}, \tag{3.146}$$

$$\mathbf{M}_0 = \int_0^h e^{\mathbf{A}(h-s)}\,\mathrm{d}s = \mathbf{A}^{-1}(e^{\mathbf{A}h} - \mathbf{I}), \tag{3.147}$$

$$\mathbf{M}_1 = \int_0^h s e^{\mathbf{A}(h-s)}\,\mathrm{d}s = \mathbf{A}^{-1}(\mathbf{M}_0 - h\mathbf{I}), \tag{3.148}$$

$$\mathbf{M}_2 = \int_0^h s^2 e^{\mathbf{A}(h-s)}\,\mathrm{d}s = \mathbf{A}^{-1}\left(2\mathbf{M}_1 - h^2\mathbf{I}\right), \tag{3.149}$$

with $\mathbf{I}$ denoting the identity matrix, provided that the inverse of matrix $\mathbf{A}$ exists. State-augmentation of (3.144) results in the $(p + 1)n$-dimensional discrete map

$$\mathbf{z}_{i+1} = \mathbf{G}_i \mathbf{z}_i \tag{3.150}$$

with $\mathbf{z}_i = (\mathbf{y}_i, \mathbf{y}_{i-1}, \ldots, \mathbf{y}_{i-p})^T$ and

$$\mathbf{G}_i = \begin{pmatrix} \mathbf{H}_{i+1}\left(e^{\mathbf{A}h} + \mathbf{F}_i\right) & \mathbf{0} & \cdots & \mathbf{0} & -\mathbf{H}_{i+1}\mathbf{F}_{i+1} & -\mathbf{H}_{i+1}\mathbf{F}_i \\ \mathbf{I} & \mathbf{0} & \cdots & \mathbf{0} & \mathbf{0} & \mathbf{0} \\ \mathbf{0} & \mathbf{I} & & \mathbf{0} & \mathbf{0} & \mathbf{0} \\ \vdots & & & & & \vdots \\ \mathbf{0} & \mathbf{0} & \cdots & \mathbf{0} & \mathbf{I} & \mathbf{0} \end{pmatrix}, \tag{3.151}$$

where $\mathbf{H}_{i+1} = (\mathbf{I} - \mathbf{F}_{i+1})^{-1}$. The approximate Floquet transition matrix is obtained as $\mathbf{\Phi} = \mathbf{G}_{p-1}\mathbf{G}_{p-2}\cdots\mathbf{G}_0$.

Higher-order versions of this approximations can also be defined; for instance,

$$\tilde{\mathbf{y}}(t) \approx \frac{(t - t_i)(t - t_{i+1})}{2h^2}\mathbf{y}_{i-1} - \frac{(t - t_{i-1})(t - t_{i+1})}{h^2}\mathbf{y}_i + \frac{(t - t_{i-1})(t - t_i)}{2h^2}\mathbf{y}_{i+1} \tag{3.152}$$

presents a second-order approximation of $\mathbf{y}(t)$ according to [66].

Chapter 4
Newtonian Examples

According to Newton's second law, the acceleration of a particle is proportional to the net force acting on it. In cases, in which the net force depends on the actual position and on the actual velocity of the particle, the system is described by a second-order ODE (due to the velocity and the acceleration being the first and the second derivatives of the position, respectively). In cases, in which the net force depends on both the actual and some delayed values of the particle's position and velocity, the system is described by a second-order DDE. Second-order systems are therefore often used in engineering to model dynamic behavior. In this chapter, some special second-order scalar DDEs are considered and analyzed by the semi-discretization method.

4.1 Parametrically Excited Delayed Oscillators

In this section, stability charts are determined for the delayed Mathieu equation

$$\ddot{x}(t) + a_1 \dot{x}(t) + (\delta + \varepsilon \cos(\omega t))\, x(t) = b_0 x(t - \tau) , \qquad (4.1)$$

where the principal period is $T = 2\pi/\omega$ and the ratio τ/T is arbitrary.

As a first step, the system is transformed into the form

$$\dot{\mathbf{x}}(t) = \mathbf{A}(t)\mathbf{x}(t) + \mathbf{B}\mathbf{u}(t - \tau) , \qquad (4.2)$$
$$\mathbf{u}(t) = \mathbf{D}\mathbf{x}(t) , \qquad (4.3)$$

where

$$\mathbf{x}(t) = \begin{pmatrix} x(t) \\ \dot{x}(t) \end{pmatrix} , \qquad \mathbf{u}(t) = \begin{pmatrix} x(t) \end{pmatrix} , \qquad (4.4)$$

$$\mathbf{A}(t) = \begin{pmatrix} 0 & 1 \\ -(\delta + \varepsilon \cos(\omega t)) & -a_1 \end{pmatrix} , \qquad \mathbf{B} = \begin{pmatrix} 0 \\ b_0 \end{pmatrix} , \qquad \mathbf{D} = \begin{pmatrix} 1 & 0 \end{pmatrix} . \qquad (4.5)$$

This system is a DDE containing a single point delay. In what follows, stability analysis is performed using the zeroth-, the improved zeroth, and the first-order semi-discretization, as shown in Section 3.3. Here, $n = 2$, $m = 1$, and $g = 1$.

Zeroth-Order Semi-Discretization

The zeroth-order approximation results in the system

$$\dot{\mathbf{y}}(t) = \mathbf{A}_i \mathbf{y}(t) + \mathbf{B} \mathbf{v}(t_{i-r}) , \quad t \in [t_i, t_{i+1}) , \tag{4.6}$$
$$\mathbf{v}(t_i) = \mathbf{D} \mathbf{y}(t_i) , \tag{4.7}$$

where $t_i = ih$ is the discrete time scale, $h = T/p$ is the discretization step, p is the period resolution, $r = \text{int}(\tau/h)$ is the delay resolution, and

$$\mathbf{A}_i = \frac{1}{h} \int_{t_i}^{t_{i+1}} \mathbf{A}(t) \, dt . \tag{4.8}$$

Solving over the interval $[t_i, t_{i+1})$ gives

$$\mathbf{y}_{i+1} = \mathbf{P}_i \mathbf{y}_i + \mathbf{R}_{i,0} \mathbf{v}_{i-r} , \tag{4.9}$$

where

$$\mathbf{P}_i = e^{\mathbf{A}_i h} , \tag{4.10}$$

$$\mathbf{R}_{i,0} = \int_0^h e^{\mathbf{A}_i (h-s)} \, ds \, \mathbf{B} . \tag{4.11}$$

If $\mathbf{A}_i^{-1}$ exists, then integration gives

$$\mathbf{R}_{i,0} = \left(e^{\mathbf{A}_i h} - \mathbf{I} \right) \mathbf{A}_i^{-1} \mathbf{B} , \tag{4.12}$$

where $\mathbf{I}$ is the 2×2 identity matrix. Equations (4.9) and (4.7) imply the $(r + 2)$-dimensional discrete map

$$\mathbf{z}_{i+1} = \mathbf{G}_i \mathbf{z}_i , \tag{4.13}$$

where

$$\mathbf{z}_i = \left(\mathbf{y}_i \ \ \mathbf{v}_{i-1} \ \ \mathbf{v}_{i-2} \ \cdots \ \mathbf{v}_{i-r} \right)^T = \left(x_i \ \ \dot{x}_i \ \ x_{i-1} \ \ x_{i-2} \ \cdots \ x_{i-r} \right)^T \tag{4.14}$$

is the augmented state vector and the coefficient matrix reads

$$\mathbf{G}_i = \begin{pmatrix} \mathbf{P}_i & \mathbf{0} & \cdots & \mathbf{0} & \mathbf{R}_{i,0} \\ \mathbf{D} & \mathbf{0} & \cdots & \mathbf{0} & \mathbf{0} \\ \mathbf{0} & \mathbf{I} & \cdots & \mathbf{0} & \mathbf{0} \\ \vdots & & & & \vdots \\ \mathbf{0} & \mathbf{0} & \cdots & \mathbf{I} & \mathbf{0} \end{pmatrix}. \tag{4.15}$$

Improved Zeroth-Order Semi-Discretization

The improved zeroth-order semi-discretization gives the approximate system

$$\dot{\mathbf{y}}(t) = \mathbf{A}_i \mathbf{y}(t) + \mathbf{B} \left(\beta_0 \mathbf{v}(t_{i-r}) + \beta_1 \mathbf{v}(t_{i-r+1}) \right) , \quad t \in [t_i, t_{i+1}) , \tag{4.16}$$

$$\mathbf{v}(t_i) = \mathbf{D}\mathbf{y}(t_i) , \tag{4.17}$$

where

$$\beta_0 = \frac{\tau + \frac{1}{2}h - rh}{h} , \quad \beta_1 = \frac{rh + \frac{1}{2}h - \tau}{h} , \tag{4.18}$$

and $r = \mathrm{int}\,(\tau/h + 1/2)$. Note that if $\tau = T$, then $r = p$ and $\beta_0 = \beta_1 = 1/2$. The solution over one discrete step is formulated as

$$\mathbf{y}_{i+1} = \mathbf{P}_i \mathbf{y}_i + \mathbf{R}_{i,0} \left(\beta_0 \mathbf{v}_{i-r} + \beta_1 \mathbf{v}_{i-r+1} \right) , \tag{4.19}$$

where $\mathbf{P}_i$ and $\mathbf{R}_{i,0}$ are given by (4.10) and (4.11), respectively. Equations (4.19) and (4.17) imply a discrete map similar to (4.13) with the discretized state $\mathbf{z}_i$ defined in (4.14), but here, the coefficient matrix reads

$$\mathbf{G}_i = \begin{pmatrix} \mathbf{P}_i & \mathbf{0} & \cdots & \mathbf{0} & \beta_1 \mathbf{R}_{i,0} & \beta_0 \mathbf{R}_{i,0} \\ \mathbf{D} & \mathbf{0} & \cdots & \mathbf{0} & \mathbf{0} & \mathbf{0} \\ \mathbf{0} & \mathbf{I} & \cdots & \mathbf{0} & \mathbf{0} & \mathbf{0} \\ \vdots & & & & & \vdots \\ \mathbf{0} & \mathbf{0} & \cdots & \mathbf{0} & \mathbf{I} & \mathbf{0} \end{pmatrix}. \tag{4.20}$$

First-Order Semi-Discretization

The first-order semi-discretization results in the approximate system

$$\dot{\mathbf{y}}(t) = \mathbf{A}_i \mathbf{y}(t) + \mathbf{B} \left(\beta_0(t) \mathbf{v}(t_{i-r}) + \beta_1(t) \mathbf{v}(t_{i-r+1}) \right) , \quad t \in [t_i, t_{i+1}) , \tag{4.21}$$

$$\mathbf{v}(t_i) = \mathbf{D}\mathbf{y}(t_i) , \tag{4.22}$$

where

$$\beta_0(t) = \frac{\tau + (i - r + 1)h - t}{h} , \quad \beta_1(t) = \frac{t - (i - r)h - \tau}{h} , \tag{4.23}$$

and $r = \mathrm{int}\,(\tau/h + 1/2)$. The solution over one discrete step can be formulated as

$$\mathbf{y}_{i+1} = \mathbf{P}_i \mathbf{y}_i + \mathbf{R}_{i,0} \mathbf{v}_{i-r} + \mathbf{R}_{i,1} \mathbf{v}_{i-r+1} \,, \tag{4.24}$$

where $\mathbf{P}_i$ is given by (4.10) and

$$\mathbf{R}_{i,0} = \int_0^h \frac{\tau - (r-1)h - s}{h} \, e^{\mathbf{A}_i(h-s)} \, ds \, \mathbf{B} \,, \tag{4.25}$$

$$\mathbf{R}_{i,1} = \int_0^h \frac{s - \tau + rh}{h} \, e^{\mathbf{A}_i(h-s)} \, ds \, \mathbf{B} \,. \tag{4.26}$$

If $\mathbf{A}_i^{-1}$ exists, then integration gives

$$\mathbf{R}_{i,0} = \left(\mathbf{A}_i^{-1} + \frac{1}{h} \left(\mathbf{A}_i^{-2} - (\tau - (r-1)h) \, \mathbf{A}_i^{-1} \right) \left(\mathbf{I} - e^{\mathbf{A}_i h} \right) \right) \mathbf{B} \,, \tag{4.27}$$

$$\mathbf{R}_{i,1} = \left(-\mathbf{A}_i^{-1} + \frac{1}{h} \left(-\mathbf{A}_i^{-2} + (\tau - rh) \, \mathbf{A}_i^{-1} \right) \left(\mathbf{I} - e^{\mathbf{A}_i h} \right) \right) \mathbf{B} \,. \tag{4.28}$$

Equations (4.24) and (4.22) imply a discrete map similar to (4.13) with the discretized state $\mathbf{z}_i$ defined in (4.14). The coefficient matrix for the first-order approximation reads

$$\mathbf{G}_i = \begin{pmatrix} \mathbf{P}_i & \mathbf{0} & \cdots & \mathbf{0} & \mathbf{R}_{i,1} & \mathbf{R}_{i,0} \\ \mathbf{D} & \mathbf{0} & \cdots & \mathbf{0} & \mathbf{0} & \mathbf{0} \\ \mathbf{0} & \mathbf{I} & \cdots & \mathbf{0} & \mathbf{0} & \mathbf{0} \\ \vdots & & & & & \vdots \\ \mathbf{0} & \mathbf{0} & \cdots & \mathbf{0} & \mathbf{I} & \mathbf{0} \end{pmatrix} \,. \tag{4.29}$$

The monodromy matrix is obtained by p repeated applications of the the discrete map $\mathbf{z}_{i+1} = \mathbf{G}_i \mathbf{z}_i$ with initial state $\mathbf{z}_0$, i.e., $\mathbf{\Phi} = \mathbf{G}_{p-1} \mathbf{G}_{p-2} \cdots \mathbf{G}_0$.

Figure 4.1 presents the stability charts for the undamped case ($a_1 = 0$) with $\varepsilon = 2$ and $\tau = T = 2\pi$ obtained by the zeroth-, the improved zeroth-, and the first-order semi-discretization methods with different period resolutions p. The diagrams were constructed by point-by-point numerical evaluation of the critical eigenvalues over a 400×200 grid of parameters δ and b_0. Stable domains associated with different approximation types are indicated by different shades of gray. The exact stability boundaries (see Section 2.4) are also given for reference. The figure shows how the boundaries for the different approximations approach the exact boundaries as the period resolution p is increasing. It can be seen that the stability boundaries obtained by the first-order semi-discretization method practically coincide with the exact boundaries already for period resolution $p = 20$. The improved zeroth-order method also provides a good approximation for $p = 20$ within the presented parameter domain. Note that in this example, the principal period T is equal to the time delay τ; consequently, the delay resolution is $r = \text{int}\,(\tau/h) = p$ for the zeroth-order method and $r = \text{int}\,(\tau/h + 1/2) = p$ for the improved zeroth-order and for the first-order method. The Matlab code for the calculation of the charts by the first-order semi-discretization method is given in Appendix A.3.

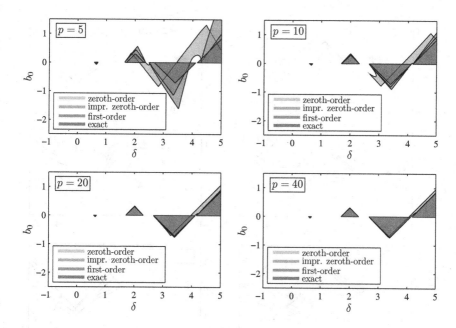

Fig. 4.1 Exact and approximate stability charts for (4.1) with $\varepsilon = 2$, $a_1 = 0$, $\tau = T = 2\pi$ and for different period resolutions p. Stable domains are indicated by gray shading.

Figure 4.2 presents the stability charts for different parameter combinations. These diagrams were determined by the first-order semi-discretization method with period resolution $p = 40$. Note that for the cases in which the time delay τ is not equal to the principal period T, the period resolution p is not equal to the delay resolution r. For instance, for $T = 4\pi$ with $\tau = 2\pi$, period resolution $p = 40$ gives the discretization step $h = \pi/10$, and the delay resolution is $r = \text{int}\,(\tau/h + 1/2) = 20$. The resolution of the parameter plane (δ, b_0) was 400×200. For the case $a_1 = 0$, stable domains are indicated by gray shading, while for cases $a_1 = 0.1$ and $a_1 = 0.2$, only the stability boundaries are presented. The undamped resonant case $a_1 = 0$, $T = \tau = 2\pi$ gives the straight stability boundaries derived in Section 2.4. Parameters $a_1 = 0$, $T = \pi$, $\tau = 2\pi$ provide also a special resonant case with straight stability boundaries. In this case, the unstable tongues of the corresponding Ince–Strutt diagram with $T = \pi$ are born at $\delta = 1, 4, 9, \ldots$, which coincide exactly with the intersection of the stability boundaries of slope $+1$ and line $b_0 = 0$ in the corresponding Hsu–Bhatt stability chart. Therefore, the stability boundaries of slope $+1$ open according to the corresponding Ince–Strutt diagram, while the boundaries of slope -1 remain fixed. For the other cases in which the ratio τ/T is not an integer, the stability charts show a complex and intricate structure that is the result of the interaction of the time-delayed term and the parametric excitation.

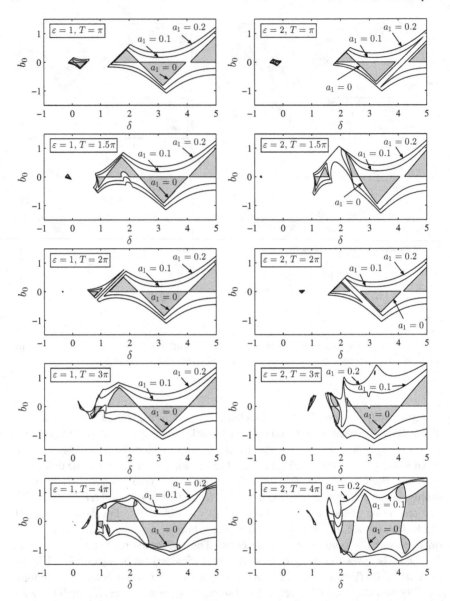

Fig. 4.2 Stability charts for (4.1) with $\tau = 2\pi$ for different values of ε, T, and a_1. The period resolution is $p = 40$ for all cases. Stable domains are indicated by gray shading.

4.2 Multiple Delays

Consider the Mathieu equation with multiple delays in the form

$$\ddot{x}(t) + a_1\dot{x}(t) + (\delta + \varepsilon\cos(\omega t))\, x(t) = \sum_{j=1}^{g} b_{0,j}\, x(t - \tau_j) + b_{1,j}\, \dot{x}(t - \tau_j) \qquad (4.30)$$

with delays $\tau_j > 0$ and principal period $T = 2\pi/\omega$. The first-order representation of this system reads

$$\dot{\mathbf{x}}(t) = \mathbf{A}(t)\mathbf{x}(t) + \sum_{j=1}^{g} \mathbf{B}_j\mathbf{u}(t - \tau_j)\,, \qquad (4.31)$$

$$\mathbf{u}(t) = \mathbf{D}\mathbf{x}(t)\,, \qquad (4.32)$$

with

$$\mathbf{x}(t) = \begin{pmatrix} x(t) \\ \dot{x}(t) \end{pmatrix}, \qquad \mathbf{u}(t) = \begin{pmatrix} x(t) \\ \dot{x}(t) \end{pmatrix}, \qquad \mathbf{D} = \begin{pmatrix} 1 & 0 \\ 0 & 1 \end{pmatrix}, \qquad (4.33)$$

$$\mathbf{A}(t) = \begin{pmatrix} 0 & 1 \\ -(\delta + \varepsilon\cos(\omega t)) & -a_1 \end{pmatrix}, \qquad \mathbf{B}_j = \begin{pmatrix} 0 & 0 \\ b_{0,j} & b_{1,j} \end{pmatrix}. \qquad (4.34)$$

Since $\mathbf{u}(t) = \mathbf{x}(t)$, the system can also be written as

$$\dot{\mathbf{x}}(t) = \mathbf{A}(t)\mathbf{x}(t) + \sum_{j=1}^{g} \mathbf{B}_j\mathbf{x}(t - \tau_j)\,. \qquad (4.35)$$

The steps for the first-order semi-discretization method are given by (3.92)–(3.101). For each particular delay τ_j, the corresponding particular delay resolution r_j is defined as $r_j = \text{int}(\tau_j/h + 1/2)$. For a fixed sampling period h, there might be cases in which some of the delays τ_j are so small that the corresponding particular delay resolutions are zero. This phenomenon typically arises when stability charts are plotted in the plane of the delay parameters including domains in which one of the delays is very small, or actually equal to zero. Here, this effect is also taken into account during the semi-discretization procedure. Assume that $r_j = 0$ for $j = 1, 2, \ldots, g_0$, $r_j = 1$ for $j = g_0 + 1, g_0 + 2, \ldots, g_0 + g_1$, and $r_j \geq 2$ for $j = g_0 + g_1 + 1, g_0 + g_1 + 2, \ldots, g$. In this case, the approximate semi-discrete system can be given as

$$\dot{\mathbf{y}}(t) = \left(\mathbf{A}_i + \sum_{j=1}^{g_0} \mathbf{B}_j \mathbf{D} \right) \mathbf{y}(t)$$

$$+ \sum_{j=g_0+1}^{g} \mathbf{B}_j \left(\beta_{j,0}(t) \mathbf{v}(t_{i-r_j}) + \beta_{j,1}(t) \mathbf{v}(t_{i-r_j+1}) \right), \quad t \in [t_i, t_{i+1}), \quad (4.36)$$

$$\mathbf{v}(t_i) = \mathbf{D} \mathbf{y}(t_i), \quad (4.37)$$

where

$$\beta_{j,0}(t) = \frac{\tau_j + (i - r_j + 1)h - t}{h}, \qquad \beta_{j,1}(t) = \frac{t - (i - r_j)h - \tau_j}{h}. \quad (4.38)$$

Thus, the small particular delays τ_j, $j = 1, 2, \ldots, g_0$, associated with zero particular delay resolution $r_j = 0$ are neglected, and the corresponding coefficient matrices $\mathbf{B}_j$ are added to $\mathbf{A}_i$, as if these terms were undelayed. If this step is not performed, then the approximated delayed terms $\mathbf{v}(t_{i-r_j+1})$ give $\mathbf{v}(t_{i+1}) = \mathbf{D} \mathbf{y}(t_{i+1})$, which should be considered when $\mathbf{y}_{i+1}$ is determined.

The solution over one discrete step can be given as

$$\mathbf{y}_{i+1} = \mathbf{P}_i \mathbf{y}_i + \sum_{j=g_0+1}^{g} \left(\mathbf{R}_{j,i,0} \mathbf{v}_{i-r_j} + \mathbf{R}_{j,i,1} \mathbf{v}_{i-r_j+1} \right), \quad (4.39)$$

where

$$\mathbf{P}_i = e^{\hat{\mathbf{A}}_i h}, \qquad \hat{\mathbf{A}}_i = \left(\mathbf{A}_i + \sum_{j=1}^{g_0} \mathbf{B}_j \mathbf{D} \right), \quad (4.40)$$

$$\mathbf{R}_{j,i,0} = \int_0^h \frac{\tau_j - (r_j - 1)h - s}{h} e^{\hat{\mathbf{A}}_i(h-s)} \, \mathrm{d}s \, \mathbf{B}_j, \quad (4.41)$$

$$\mathbf{R}_{j,i,1} = \int_0^h \frac{s - \tau_j + r_j h}{h} e^{\hat{\mathbf{A}}_i(h-s)} \, \mathrm{d}s \, \mathbf{B}_j. \quad (4.42)$$

If $\mathbf{A}_i^{-1}$ exists, then integration gives

$$\mathbf{R}_{j,i,0} = \left(\hat{\mathbf{A}}_i^{-1} + \frac{1}{h} \left(\hat{\mathbf{A}}_i^{-2} - (\tau_j - (r_j - 1)h) \hat{\mathbf{A}}_i^{-1} \right) \left(\mathbf{I} - e^{\hat{\mathbf{A}}_i h} \right) \right) \mathbf{B}_j, \quad (4.43)$$

$$\mathbf{R}_{j,i,1} = \left(-\hat{\mathbf{A}}_i^{-1} + \frac{1}{h} \left(-\hat{\mathbf{A}}_i^{-2} + (\tau_j - r_j h) \hat{\mathbf{A}}_i^{-1} \right) \left(\mathbf{I} - e^{\hat{\mathbf{A}}_i h} \right) \right) \mathbf{B}_j. \quad (4.44)$$

Equations (4.39) and (4.37) imply the discrete map

$$\mathbf{z}_{i+1} = \mathbf{G}_i \mathbf{z}_i, \quad (4.45)$$

where

$$\mathbf{z}_i = \left(\mathbf{y}_i \ \ \mathbf{v}_{i-1} \ \ \mathbf{v}_{i-2} \ \ \cdots \ \ \mathbf{v}_{i-r} \right)^T = \left(x_i \ \ \dot{x}_i \ \ x_{i-1} \ \ \dot{x}_{i-1} \ \ \cdots \ \ x_{i-r} \ \ \dot{x}_{i-r} \right)^T \quad (4.46)$$

is the augmented state vector and the coefficient matrix reads

$$
G_i = \begin{pmatrix} P_i & 0 & \cdots & 0 & 0 \\ \hline D & 0 & \cdots & 0 & 0 \\ 0 & I & \cdots & 0 & 0 \\ \vdots & & \ddots & & \vdots \\ 0 & 0 & \cdots & I & 0 \end{pmatrix} + \sum_{j=g_0+1}^{g_0+g_1} \left(\begin{array}{c|ccc} R_{j,i,1}D & R_{j,i,0} & 0 & \cdots & 0 \\ \hline 0 & 0 & 0 & \cdots & 0 \\ 0 & 0 & 0 & \cdots & 0 \\ \vdots & \vdots & & & \vdots \\ 0 & 0 & 0 & \cdots & 0 \end{array} \right)
$$

$$
+ \sum_{j=g_0+g_1+1}^{g} \overset{\displaystyle \quad 1 \qquad\qquad\quad r_j-1 \quad r_j \qquad\qquad r}{\begin{pmatrix} 0 & 0 & \cdots & 0 & R_{j,i,1} & R_{j,i,0} & 0 & \cdots & 0 \\ \hline 0 & 0 & \cdots & 0 & 0 & 0 & 0 & \cdots & 0 \\ 0 & 0 & \cdots & 0 & 0 & 0 & 0 & \cdots & 0 \\ \vdots & \vdots & & \vdots & \vdots & \vdots & \vdots & & \vdots \\ 0 & 0 & \cdots & 0 & 0 & 0 & 0 & \cdots & 0 \end{pmatrix}} .
$$

$$(4.47)$$

If $r_j = 1$, then the corresponding term in (4.39) is

$$
R_{j,i,0}v_{i-1} + R_{j,i,1}v_i = R_{j,i,0}v_{i-1} + R_{j,i,1}Dy_i . \tag{4.48}
$$

Therefore, matrix $R_{j,i,1}D$ appears in the upper left block for $j = g_0+1, g_0+2, \ldots, g_0+ g_1$. For $j = g_0 + g_1 + 1, g_0 + g_1 + 2, \ldots, g$, the corresponding matrices $R_{j,i,1}$ and $R_{j,i,0}$ are located at the $(r_j - 1)$th and r_jth places in the upper right block of G_i, respectively. The dimension of the augmented system is $(2r+2)$, where $r = \max(r_j)$, $j = 1, 2, \ldots, g$, or, what is equivalent, $r = \mathrm{int}(\tau_{max}/h + 1/2)$ with $\tau_{max} = \max(\tau_j)$, $j = 1, 2, \ldots, g$.

The monodromy matrix is obtained by p repeated applications of (4.45) with initial state z_0, giving $\Phi = G_{p-1}G_{p-2} \cdots G_0$. Stability properties can be determined by the analysis of the eigenvalues of Φ. A sample Matlab code for the calculation of stability charts for (4.49) is given in Appendix A.4.

Three different equations are considered as case studies. The first example reads

$$
\ddot{x}(t) + \left(6 + \varepsilon \cos\left(\tfrac{2\pi}{5}t\right)\right) x(t) = x(t - \tau_1) + x(t - \tau_2) . \tag{4.49}
$$

This equation corresponds to (4.30) with $g = 2$, $a_1 = 0$, $\delta = 6$, $b_{0,1} = 1$, $b_{0,2} = 1$, $b_{1,1} = 0$, $b_{1,2} = 0$, and $T = 2\pi/\omega = 5$. The stability charts in the plane of the delay parameters (τ_1, τ_2) are presented in Figure 4.3 for different amplitudes ε of the parametric forcing. The period resolution was $p = 50$, and the maximum delay resolution was $r = 100$, which corresponds to the maximum delay $\tau_{max} = 10$ in Figure 4.3. Note that when the stability charts are plotted in the plane of the delay parameters, then a relatively large delay resolution should be applied, since both small and large delays may appear at different parts of the stability chart. For instance, the parameter pair $(\tau_1, \tau_2) = (0.1, 10)$ is associated with the particular delay resolutions $r_1 = 1$ and $r_2 = 100$, respectively. The resolution of the parameter plane (τ_1, τ_2) was 200×200, which is commensurate with the delay resolution $r = 100$. Since the coefficients of the delayed terms in (4.49) are equal, the stability chart is symmetric

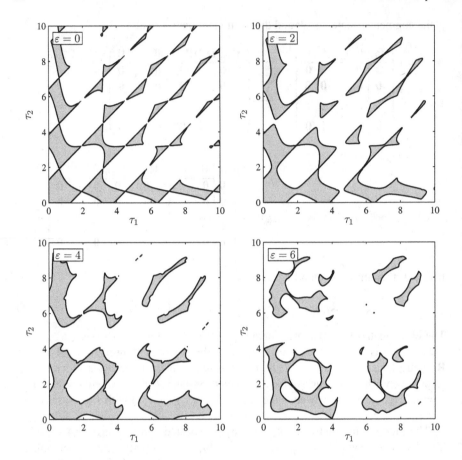

Fig. 4.3 Stability charts for (4.49) with stable domains indicated by gray shading.

with respect to the $\tau_1 = \tau_2$ axis. The time-invariant case ($\varepsilon = 0$) for (4.49) was analyzed, and the corresponding stability chart was presented by Stepan [255].

The second example is the equation

$$\ddot{x}(t) + 1.5\dot{x}(t) + \big(60 + \varepsilon \cos(2\pi t)\big)x(t) = -14x(t - \tau_1) - 1.4\dot{x}(t - \tau_2) . \qquad (4.50)$$

This equation corresponds to (4.30) with $g = 2$, $a_1 = 1.5$, $\delta = 60$, $b_{0,1} = -14$, $b_{0,2} = 0$, $b_{1,1} = 0$, $b_{1,2} = -1.4$, and $T = 1$. The stability charts in the plane (τ_1, τ_2) are presented in Figure 4.4 for different amplitudes ε. Here, the period resolution was $p = 50$ and the maximum delay resolution was $r = 100$. The resolution of the parameter plane (τ_1, τ_2) was 200×200. The time-invariant case ($\varepsilon = 0$) for (4.50) with slightly different parameters was analyzed and the corresponding stability chart was presented by Olgac et al. [212].

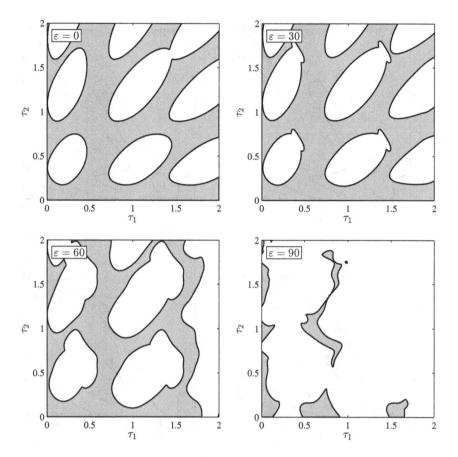

Fig. 4.4 Stability charts for (4.50) with stable domains indicated by gray shading.

The third example equation reads

$$\ddot{x}(t) + 3.5\dot{x}(t) + \left(8 + \varepsilon \cos(0.8\pi t)\right)x(t) = -0.5x(t - \tau_1) - 2\dot{x}(t - \tau_1)$$
$$- 8x(t - \tau_2) + \dot{x}(t - \tau_2) - 5x(t - \tau_1 - \tau_2) . \quad (4.51)$$

This equation corresponds to (4.30) with $g = 3$, $a_1 = 3.5$, $\delta = 8$, $b_{0,1} = -0.5$, $b_{0,2} = -8$, $b_{0,3} = -5$, $b_{1,1} = -2$, $b_{1,2} = 1$, $b_{1,3} = 0$, $\tau_3 = \tau_1 + \tau_2$, and $T = 2.5$. The stability charts in the plane (τ_1, τ_2) are presented in Figure 4.5 for different amplitudes ε. The period resolution was $p = 50$, and the maximum delay resolution was $r = 200$. Note that the maximum delay in Figure 4.5 is $\tau_{max} = \tau_{3,max} = \tau_{1,max} + \tau_{2,max} = 20$. The resolution of the parameter plane (τ_1, τ_2) was 200×200. The stability chart for the time-invariant case ($\varepsilon = 0$) of (4.51) was presented by Sipahi and Olgac [250].

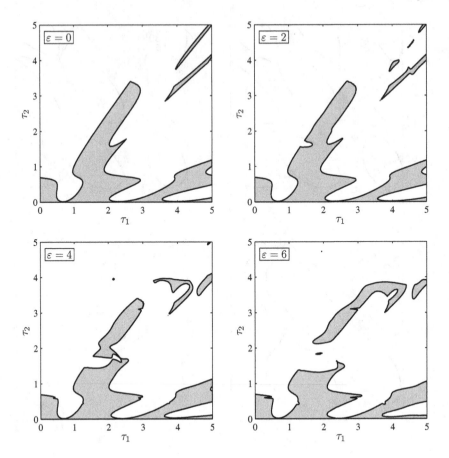

Fig. 4.5 Stability charts for (4.51) with stable domains indicated by gray shading.

4.3 Distributed Delays

In this section, stability charts are determined for the Mathieu equation with distributed delay in the form

$$\ddot{x}(t) + a_1 \dot{x}(t) + (\delta + \varepsilon \cos(\omega t)) \, x(t) = b_0 \int_{-\sigma}^{0} w(\vartheta) x(t + \vartheta) \, \mathrm{d}\vartheta \,. \tag{4.52}$$

Here, the delay effect is described by the weight function $w(\vartheta)$, $\vartheta \in [-\sigma, 0]$. The first-order representation of this system reads

$$\dot{\mathbf{x}}(t) = \mathbf{A}(t)\mathbf{x}(t) + \int_{-\sigma}^{0} \mathbf{W}(\vartheta)\mathbf{u}(t + \vartheta)\,\mathrm{d}\vartheta\,, \tag{4.53}$$

$$\mathbf{u}(t) = \mathbf{D}\mathbf{x}(t)\,, \tag{4.54}$$

where

$$\mathbf{x}(t) = \begin{pmatrix} x(t) \\ \dot{x}(t) \end{pmatrix}, \qquad \mathbf{u}(t) = \begin{pmatrix} x(t) \end{pmatrix}, \qquad \mathbf{D} = \begin{pmatrix} 1 & 0 \end{pmatrix}, \tag{4.55}$$

$$\mathbf{A}(t) = \begin{pmatrix} 0 & 1 \\ -(\delta + \varepsilon\cos(\omega t)) & -a_1 \end{pmatrix}, \qquad \mathbf{W}(\vartheta) = \begin{pmatrix} 0 \\ b_0 w(\vartheta) \end{pmatrix}. \tag{4.56}$$

The first step of the semi-discretization method is that the distributed delay is approximated by a linear combination of point delay terms according to (3.40). This step corresponds to the approximation of the weight function by shifted Dirac delta distributions, as shown in Figure 4.6. The second step is the approximation of the time-periodic coefficients by their averages. The resulting system is therefore a DDE with point delays and constant coefficients in the form

$$\dot{\tilde{\mathbf{x}}}(t) = \mathbf{A}_i\tilde{\mathbf{x}}(t) + \sum_{k=1}^{f} \tilde{\mathbf{W}}_k\tilde{\mathbf{u}}(t - \tilde{\tau}_k)\,, \qquad t \in [t_i, t_{i+1})\,, \tag{4.57}$$

$$\tilde{\mathbf{u}}(t) = \mathbf{D}\tilde{\mathbf{x}}(t)\,, \tag{4.58}$$

where $t_i = ih$, $h = T/p$, with p being the period resolution, $\tilde{\tau}_k = (k - 1/2)h$, $k = 1, 2, \ldots, f$, $f = \mathrm{ceil}(\sigma/h)$, with ceil being the ceiling function, and

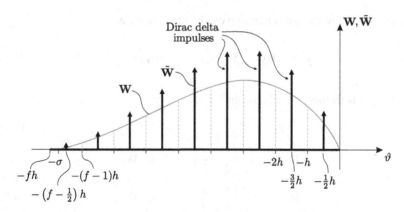

Fig. 4.6 Approximation of the weight function of the distributed delay by shifted Dirac delta distributions.

$$\mathbf{A}_i = \frac{1}{h} \int_{t_i}^{t_{i+1}} \mathbf{A}(t)\, dt \,, \tag{4.59}$$

$$\tilde{\mathbf{W}}_k = \int_{-kh}^{-(k-1)h} \mathbf{W}(\vartheta)\, d\vartheta \,, \qquad k = 1, 2, \ldots, f-1 \,, \tag{4.60}$$

$$\tilde{\mathbf{W}}_f = \int_{-\sigma}^{-(f-1)h} \mathbf{W}(\vartheta)\, d\vartheta \,. \tag{4.61}$$

The final step of the semi-discretization is the approximation of the delayed terms by piecewise constant (zeroth-order and improved zeroth-order) or by piecewise linear (first-order) ones. Here, the first-order approximation will be used, for which the approximate equation reads

$$\dot{\mathbf{y}}(t) = \mathbf{A}_i \mathbf{y}(t) + \sum_{k=1}^{f} \tilde{\mathbf{W}}_k \left(\beta_{k,0}(t) \mathbf{v}(t_{i-r_k}) + \beta_{k,1}(t) \mathbf{v}(t_{i-r_k+1}) \right), \quad t \in [t_i, t_{i+1}) \,, \tag{4.62}$$

$$\mathbf{v}(t_i) = \mathbf{D}\mathbf{y}(t_i) \,, \tag{4.63}$$

where

$$r_k = \text{int}\left(\frac{\tau_k}{h} + \frac{1}{2} \right) = \text{int}\left(k - \frac{1}{2} + \frac{1}{2} \right) = k \,, \tag{4.64}$$

$$\beta_{k,0}(t) = \frac{\tau_k + (i - r_k + 1)h - t}{h} = \frac{\left(i + \frac{1}{2} \right)h - t}{h} \,, \tag{4.65}$$

$$\beta_{k,1}(t) = \frac{t - (i - r_k)h - \tau_k}{h} = \frac{t - \left(i - \frac{1}{2} \right)h}{h} \,. \tag{4.66}$$

The solution over one discrete step can be formulated as

$$\mathbf{y}_{i+1} = \mathbf{P}_i \mathbf{y}_i + \sum_{k=1}^{r} \mathbf{R}_{i,k,0} \mathbf{v}_{i-k} + \mathbf{R}_{i,k,1} \mathbf{v}_{i-k+1} \,, \tag{4.67}$$

where $r = f$ is the delay resolution and

$$\mathbf{P}_i = e^{\mathbf{A}_i h} \,, \tag{4.68}$$

$$\mathbf{R}_{i,k,0} = \int_0^h \left(\frac{1}{2} - \frac{s}{h} \right) e^{\mathbf{A}_i(h-s)}\, ds\, \tilde{\mathbf{W}}_k \,, \tag{4.69}$$

$$\mathbf{R}_{i,k,1} = \int_0^h \left(\frac{s}{h} + \frac{1}{2} \right) e^{\mathbf{A}_i(h-s)}\, ds\, \tilde{\mathbf{W}}_k \,. \tag{4.70}$$

If $\mathbf{A}_i^{-1}$ exists, then integration gives

$$\mathbf{R}_{i,k,0} = \left(\mathbf{A}_i^{-1} + \left(\frac{1}{2}\mathbf{A}_i^{-1} - \frac{1}{h}\mathbf{A}_i^{-2}\right)\left(e^{\mathbf{A}_i h} - \mathbf{I}\right)\right)\tilde{\mathbf{W}}_k , \tag{4.71}$$

$$\mathbf{R}_{i,k,1} = \left(-\mathbf{A}_i^{-1} + \left(\frac{1}{2}\mathbf{A}_i^{-1} + \frac{1}{h}\mathbf{A}_i^{-2}\right)\left(e^{\mathbf{A}_i h} - \mathbf{I}\right)\right)\tilde{\mathbf{W}}_k . \tag{4.72}$$

Equations (4.67) and (4.63) imply the $(r + 2)$-dimensional discrete map

$$\mathbf{z}_{i+1} = \mathbf{G}_i \mathbf{z}_i , \tag{4.73}$$

where

$$\mathbf{z}_i = \begin{pmatrix} \mathbf{y}_i & \mathbf{v}_{i-1} & \mathbf{v}_{i-2} & \cdots & \mathbf{v}_{i-r} \end{pmatrix}^T = \begin{pmatrix} x_i & \dot{x}_i & x_{i-1} & x_{i-2} & \cdots & x_{i-r} \end{pmatrix}^T \tag{4.74}$$

is the augmented state vector and the coefficient matrix reads

$$\mathbf{G}_i = \begin{pmatrix} \mathbf{P}_i + \mathbf{R}_{i,1,1}\mathbf{D} & \mathbf{R}_{i,2,1} + \mathbf{R}_{i,1,0} & \mathbf{R}_{i,3,1} + \mathbf{R}_{i,2,0} & \cdots & \mathbf{R}_{i,r,1} + \mathbf{R}_{i,r-1,0} & \mathbf{R}_{i,r,0} \\ \mathbf{D} & \mathbf{0} & \mathbf{0} & \cdots & \mathbf{0} & \mathbf{0} \\ \mathbf{0} & \mathbf{I} & \mathbf{0} & \cdots & \mathbf{0} & \mathbf{0} \\ \vdots & & & & \vdots & \\ \mathbf{0} & \mathbf{0} & \mathbf{0} & \cdots & \mathbf{I} & \mathbf{0} \end{pmatrix}. \tag{4.75}$$

The monodromy matrix is obtained by p repeated applications of (4.73) with initial state $\mathbf{z}_0$, giving $\boldsymbol{\Phi} = \mathbf{G}_{p-1}\mathbf{G}_{p-2}\cdots\mathbf{G}_0$.

Three different delay distributions are analyzed:

$$w(\vartheta) = 1 , \tag{4.76}$$

$$w(\vartheta) = \frac{\pi}{2}\sin(\pi\vartheta) , \tag{4.77}$$

$$w(\vartheta) = \frac{\pi}{2}\sin(\pi\vartheta) + \frac{13\pi}{77}\sin(2\pi\vartheta) , \tag{4.78}$$

with $\vartheta \in [-1, 0]$, i.e., $\sigma = 1$. The graphs of these kernel functions are presented in Figure 4.7. The time-invariant cases ($\varepsilon = 0$) for these distributions were analyzed by Stepan in [255] and [256] using his criteria (1.25) for the number of unstable

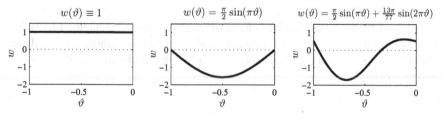

Fig. 4.7 Graphs of kernel functions (4.76), (4.77), and (4.78).

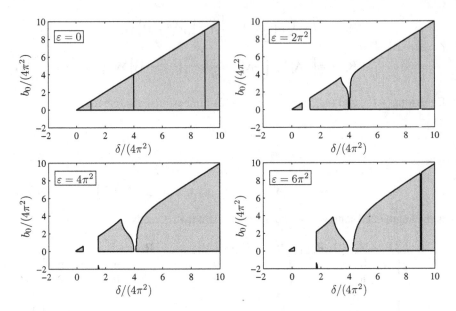

Fig. 4.8 Stability charts of (4.52) with $w(\vartheta) \equiv 1$, $\sigma = 1$, $T = 2\pi/\omega = 1/2$, $a_1 = 0$. Stable domains are indicated by gray shading.

characteristic exponents. Here, these examples are considered with parametric excitation when $\varepsilon \neq 0$. Stability diagrams in Figures 4.8, 4.9, and 4.10 were constructed by point-by-point numerical evaluation of the critical eigenvalues over a 300×200 grid of parameters δ and b_0. For the calculations, the period resolution was $p = 20$ and the delay resolution was $r = 40$. The Matlab code for the calculation of the charts is given in Appendix A.5.

Figure 4.8 presents the stability charts of (4.52) with kernel function (4.76), principal period $T = 2\pi/\omega = 1/2$, and damping parameter $a_1 = 0$ for different amplitudes ε of the parametric forcing. The stability diagram for the time-invariant case ($\varepsilon = 0$) can be found in [256] for reference.

Stability charts of (4.52) with kernel function (4.77) can be seen in Figure 4.9 for $T = 2\pi/\omega = 1/2$, $a_1 = 0$, and for different values of ε. The stability chart and the number of unstable characteristic exponents for the time-invariant case ($\varepsilon = 0$) were given in [255].

Figure 4.10 shows the stability charts of (4.52) with kernel function (4.78) for $T = 2\pi/\omega = 1/2$, $a_1 = 0$, and for different values of ε. The stability chart and the number of unstable characteristic exponents for the time-invariant case ($\varepsilon = 0$) were provided in [255].

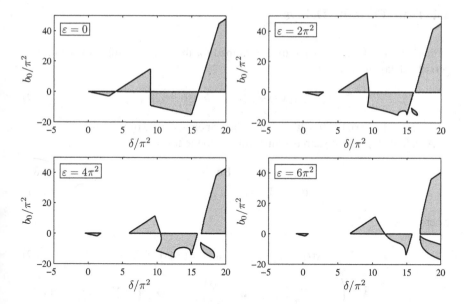

Fig. 4.9 Stability charts of (4.52) with $w(\vartheta) = \frac{\pi}{2}\sin(\pi\theta)$, $\sigma = 1$, $T = 2\pi/\omega = 1/2$, $a_1 = 0$. Stable domains are indicated by gray shading.

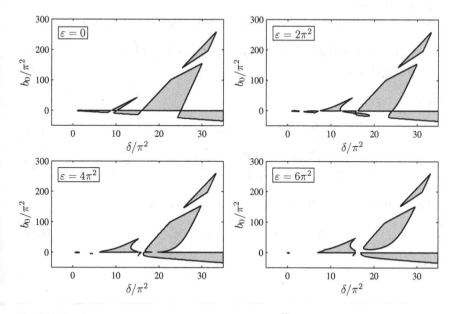

Fig. 4.10 Stability charts of (4.52) with $w(\vartheta) = \frac{\pi}{2}\sin(\pi\theta) + \frac{13\pi}{77}\sin(2\pi\vartheta)$, $\sigma = 1$, $T = 2\pi/\omega = 1/2$, $a_1 = 0$. Stable domains are indicated by gray shading.

4.4 Time-Periodic Delays

In this section, stability charts are determined for the delayed oscillator with time-periodic delay in the form

$$\ddot{x}(t) + a_1 \dot{x}(t) + a_0 x(t) = b_0 x(t - \tau(t)) , \tag{4.79}$$

where $\tau(t + T) = \tau(t) > 0$, with T being the principal period of the system.

As a first step, the system is transformed into the form

$$\dot{\mathbf{x}}(t) = \mathbf{A}\mathbf{x}(t) + \mathbf{B}\mathbf{u}(t - \tau(t)) , \tag{4.80}$$

$$\mathbf{u}(t) = \mathbf{D}\mathbf{x}(t) , \tag{4.81}$$

where

$$\mathbf{x}(t) = \begin{pmatrix} x(t) \\ \dot{x}(t) \end{pmatrix} , \qquad \mathbf{u}(t) = \begin{pmatrix} x(t) \end{pmatrix} , \tag{4.82}$$

$$\mathbf{A} = \begin{pmatrix} 0 & 1 \\ -a_0 & -a_1 \end{pmatrix} , \qquad \mathbf{B} = \begin{pmatrix} 0 \\ b_0 \end{pmatrix} , \qquad \mathbf{D} = \begin{pmatrix} 1 & 0 \end{pmatrix} . \tag{4.83}$$

This system is a DDE containing a time-varying point delay. Stability analysis is performed using the first-order semi-discretization, as shown in Section 3.3. Here, $n = 2, m = 1$, and $g = 1$.

The first-order semi-discretization results in the approximate system

$$\dot{\mathbf{y}}(t) = \mathbf{A}\mathbf{y}(t) + \mathbf{B}\left(\beta_{i,0}(t)\mathbf{v}(t_{i-r_i}) + \beta_{i,1}(t)\mathbf{v}(t_{i-r_i+1})\right) , \quad t \in [t_i, t_{i+1}) , \tag{4.84}$$

$$\mathbf{v}(t_i) = \mathbf{D}\mathbf{y}(t_i) , \tag{4.85}$$

where

$$\beta_{i,0}(t) = \frac{\tau_i + (i - r_i + 1)h - t}{h} , \qquad \beta_{i,1}(t) = \frac{t - (i - r_i)h - \tau_i}{h} , \tag{4.86}$$

$$r_i = \text{int}\left(\frac{\tau_i}{h} + \frac{1}{2}\right) , \qquad \tau_i = \frac{1}{h} \int_{t_i}^{t_{i+1}} \tau(t) \, dt , \qquad i = 1, 2, \dots, p . \tag{4.87}$$

The solution over one discrete step can be formulated as

$$\mathbf{y}_{i+1} = \mathbf{P}\mathbf{y}_i + \mathbf{R}_{i,0}\mathbf{v}_{i-r_i} + \mathbf{R}_{i,1}\mathbf{v}_{i-r_i+1} , \tag{4.88}$$

where

$$\mathbf{P} = e^{\mathbf{A}h} \, , \tag{4.89}$$

$$\mathbf{R}_{i,0} = \int_0^h \frac{\tau_i - (r_i - 1)h - s}{h} \, e^{\mathbf{A}(h-s)} \, ds \, \mathbf{B} \, , \tag{4.90}$$

$$\mathbf{R}_{i,1} = \int_0^h \frac{s - \tau_i + r_i h}{h} \, e^{\mathbf{A}(h-s)} \, ds \, \mathbf{B} \, . \tag{4.91}$$

If $\mathbf{A}^{-1}$ exists, then integration gives

$$\mathbf{R}_{i,0} = \left(\mathbf{A}^{-1} + \frac{1}{h} \left(\mathbf{A}^{-2} - (\tau_i - (r_i - 1)h) \, \mathbf{A}^{-1} \right) \left(\mathbf{I} - e^{\mathbf{A}h} \right) \right) \mathbf{B} \, , \tag{4.92}$$

$$\mathbf{R}_{i,1} = \left(-\mathbf{A}^{-1} + \frac{1}{h} \left(-\mathbf{A}^{-2} + (\tau_i - r_i h) \, \mathbf{A}^{-1} \right) \left(\mathbf{I} - e^{\mathbf{A}h} \right) \right) \mathbf{B} \, . \tag{4.93}$$

Equations (4.88) and (4.85) imply the discrete map

$$\mathbf{z}_{i+1} = \mathbf{G}_i \mathbf{z}_i \, , \tag{4.94}$$

where

$$\mathbf{z}_i = \begin{pmatrix} \mathbf{y}_i & \mathbf{v}_{i-1} & \mathbf{v}_{i-2} & \cdots & \mathbf{v}_{i-r} \end{pmatrix}^T = \begin{pmatrix} x_i & \dot{x}_i & x_{i-1} & x_{i-2} & \cdots & x_{i-r} \end{pmatrix}^T \tag{4.95}$$

is the augmented state vector and the coefficient matrix reads

$$\mathbf{G}_i = \begin{array}{c} \quad {\scriptstyle 1} \qquad\qquad {\scriptstyle r_i-1 \;\; r_i} \qquad\qquad {\scriptstyle r} \\ \left(\begin{array}{c|ccccccccc} \mathbf{P} & \mathbf{0} & \cdots & \mathbf{0} & \mathbf{R}_{i,1} & \mathbf{R}_{i,0} & \mathbf{0} & \cdots & \mathbf{0} & \mathbf{0} \\ \hline \mathbf{D} & \mathbf{0} & \cdots & \mathbf{0} & \mathbf{0} & \mathbf{0} & \mathbf{0} & \cdots & \mathbf{0} & \mathbf{0} \\ \mathbf{0} & \mathbf{I} & \cdots & \mathbf{0} & \mathbf{0} & \mathbf{0} & \mathbf{0} & \cdots & \mathbf{0} & \mathbf{0} \\ \vdots & & & & & & & & & \vdots \\ \mathbf{0} & \mathbf{0} & \cdots & \mathbf{0} & \mathbf{0} & \mathbf{0} & \mathbf{0} & \cdots & \mathbf{I} & \mathbf{0} \end{array} \right) \end{array} . \tag{4.96}$$

Here, $r = \max(r_i)$, $i = 1, 2, \ldots, p$. Matrices $\mathbf{R}_{i,1}$ and $\mathbf{R}_{i,0}$ are located at the $(r_i - 1)$th and r_ith place in the upper right block of $\mathbf{G}_i$, respectively. The monodromy matrix is obtained by p repeated applications of the discrete map $\mathbf{z}_{i+1} = \mathbf{G}_i \mathbf{z}_i$ with initial state $\mathbf{z}_0$, i.e., $\mathbf{\Phi} = \mathbf{G}_{p-1} \mathbf{G}_{p-2} \cdots \mathbf{G}_0$.

Figure 4.11 shows the stability charts of (4.79) with $\tau(t) = 2\pi(1 + \varepsilon \cos(2\pi t/T))$ for different values of ε, T, and a_1. For the case $a_1 = 0$, stable domains are indicated by gray shading, while for cases $a_1 = 0.1$ and $a_1 = 0.2$, only the stability boundaries are presented. The charts were constructed by point-by-point numerical evaluation of the critical eigenvalues over a 400×200 grid of parameters a_0 and b_0. The period resolution was set to $p = 20T/\pi$ such that the discretization step was $h = \pi/20$ for all cases. The corresponding delay resolution was $r = 48$ for $\varepsilon = 0.2$ and $r = 56$ for $\varepsilon = 0.4$. The Matlab code for the calculation of the charts is given in Appendix A.6.

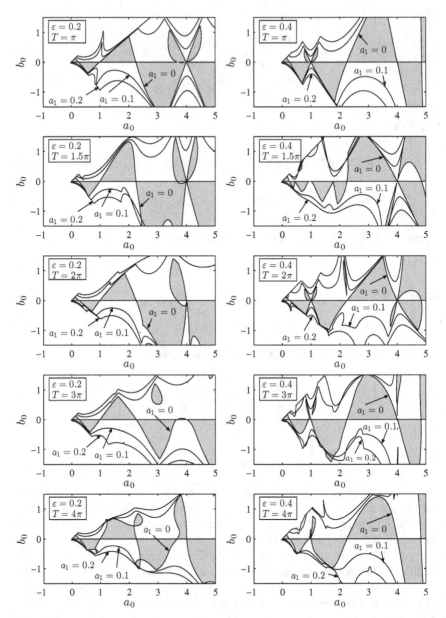

Fig. 4.11 Stability charts for (4.79) with $\tau(t) = 2\pi(1 + \varepsilon \cos(2\pi t/T))$. Stable domains are indicated by gray shading.

Chapter 5
Engineering Applications

Time-delay systems appear in several engineering problems, such as wheel shimmy [230, 257, 273, 274], car-following traffic models [213, 214, 215], feedback stabilization problems [194, 284, 252, 113], and machine tool chatter [280, 281, 255, 5]. In this chapter, engineering models are considered in which the time delay is coupled with parametric forcing. The first example to be discussed is the turning process with varying spindle speed, which is described by a DDE with time-periodic delay. Then, milling processes are considered for which time-periodic coefficients arise in the model equations due to the tooth-passing excitation of the rotating tool. After that, the act-and-wait control concept with periodically varying control gains is analyzed through an application to the stick-balancing problem with reflex delay and to a force-control process. Finally, the stick-balancing model with reflex delay is investigated in the case of parametric forcing at the stick's base.

5.1 Varying Spindle Speed Turning

The history of machine tool chatter goes back to a century, when Taylor [277] described machine tool chatter as the "most obscure and delicate of all problems facing the machinist." After the extensive work of Tobias [282, 281], Tlusty et al. [280], and Kudinov [161], the so-called regenerative effect became the most commonly accepted explanation for machine tool chatter [5, 279, 231]. This effect is related to the cutting-force variation due to the wavy workpiece surface cut one revolution ago. The phenomenon can be described by involving time delay in the model equations. Stability properties of the machining process are depicted by the so-called stability lobe diagrams, which plot the maximum stable axial depths of cut versus the spindle speed. These diagrams provide a guide to the machinist to select the optimal technological parameters in order to achieve maximum material removal rate without chatter [254, 233, 25]. There are different techniques to suppress machine tool chatter, such as the application of a vibration absorber [276, 248, 197, 297] and impedance modulation [234].

The idea of suppressing chatter by spindle speed variation came up first in the 1970s, when a great deal of effort was focused in this area [115, 275, 110, 239, 238]. The governing equation for cutting with time-varying spindle speed is a DDE with time-varying delay. There are several approaches to determining the stability properties of these processes. Sexton et al. [239] approximated the quasiperiodic solutions of the system by periodic ones and applied the harmonic balance method to derive stability boundaries. Pakdemirli and Ulsoy [217] used angle coordinate as an independent variable instead of time, following Tsao et al. [283], and obtained a DDE with constant time delay and with periodic coefficients. They used the perturbation technique called the method of strained parameters for stability analysis. Jayaram et al. [141] used quasiperiodic trial solutions for the system, combined the Fourier expansion with an expansion with respect to Bessel functions, and determined stability boundaries by the harmonic balance method. Namachchivaya and Beddini [204] transformed the time dependency from the delay term to the coefficients, and also carried out some nonlinear analysis using the small-perturbations technique. The full-discretization technique was used by Sastry et al. [228] and by Wu et al. [296] for sinusoidal speed modulation and by Yilmaz et al. [300] for random spindle speed modulation. The mathematical background for the full-discretization of DDEs with varying delays (also for systems with state-dependent delays) was presented by Győri et al. [93, 94]. The semi-discretization method was applied to the problem by Insperger and Stepan [125] and by Insperger [117] for sinusoidal and for piecewise linear (sawtooth-like) spindle speed modulations.

5.1.1 Mechanical Model

Figure 5.1 shows the chip removal process in an orthogonal turning operation for an ideally rigid tool and for a compliant tool. In the latter case, the tool experiences bending vibrations in directions x and y and leaves a wavy surface behind. The system can be modeled as a two-degrees-of-freedom oscillator excited by the cutting force, as shown in Figure 5.2. If there is no dynamic coupling between the x and y directions, then the governing equation can be given as

$$m\ddot{x}(t) + c_x\dot{x}(t) + k_x x(t) = F_x(t) , \tag{5.1}$$

$$m\ddot{y}(t) + c_y\dot{y}(t) + k_y y(t) = F_y(t) , \tag{5.2}$$

where m, c_x, c_y, k_x, and k_y are the modal mass and the damping and stiffness parameters in the x and y directions, respectively. The cutting force is given in the form

$$F_x(t) = K_x w h^q(t) , \tag{5.3}$$

$$F_y(t) = K_y w h^q(t) , \tag{5.4}$$

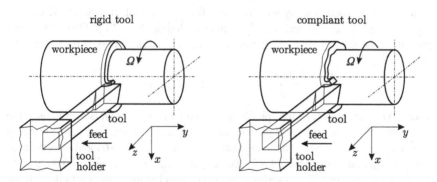

Fig. 5.1 Chip removal in orthogonal turning processes in the case of an ideally rigid tool and real compliant tool.

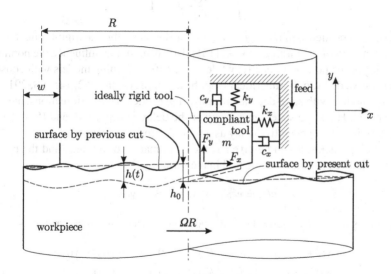

Fig. 5.2 Surface regeneration in an orthogonal turning process.

where K_x and K_y are the cutting-force coefficients in the tangential (x) and the normal (y) directions, w is the depth of cut (also known as the width of cut or the chip width in cases of orthogonal cutting), $h(t)$ is the instantaneous chip thickness, and q is the cutting-force exponent. Note that other formulas for the cutting force are also used in the literature; see, e.g., [149, 242, 68]. In this model, it is assumed that the tool never leaves the workpiece, that is, $h(t) > 0$ during the cutting process.

If the tool were rigid, then the chip thickness would be constant $h(t) \equiv h_0$, which is just equal to the feed per revolution. However, in reality, the tool experiences vibrations that are recorded on the workpiece, and after one revolution, the tool cuts this wavy surface. The chip thickness $h(t)$ is determined by the feed motion, by the current tool position, and by an earlier position of the tool. It is assumed that the spindle speed is varying periodically in time such that $\Omega(t + T) = \Omega(t) > 0$. In this case, the time delay τ between the present and the previous cuts is determined by the equation

$$\int_{t-\tau(t)}^{t} \frac{R\, 2\pi\, \Omega(s)}{60}\, \mathrm{d}s = 2R\pi + x(t) - x(t - \tau(t)) , \tag{5.5}$$

where $\Omega(t)$ is the time-varying spindle speed given in [rpm] and R is the radius of the workpiece. Equation (5.5) is in fact an implicit equation for the time delay. It can be seen that the delay actually depends also on the current state $x(t)$ and on a delayed state $x(t-\tau)$, that is, the time delay is state-dependent: $\tau(t, x_t)$, where $x_t(\vartheta) = x(t+\vartheta)$, $\vartheta \in [-\sigma, 0]$, with $\sigma \in \mathbb{R}^+$ describing the maximum length of the past effect. If the displacement $x(t)$ is negligible compared to the radius R of the workpiece, then (5.5) gives

$$\int_{t-\tau(t)}^{t} \frac{\Omega(s)}{60}\, \mathrm{d}s = 1 . \tag{5.6}$$

The state-dependency of the regenerative delay has actually a negligible effect on the linear stability of the turning process [131]. However, it may influence the nonlinear behavior of the process radically; while in the case of turning models with constant regenerative delay, only subcritical Hopf bifurcations occur [262, 145, 68, 291], for turning models with state-dependent delay, supercritical Hopf bifurcations may also arise [132]. Here, the state-dependency of the delay is neglected, and (5.6) is used to determine the regenerative delay as a time-dependent function only.

The chip thickness can be given as the combination of the feed and the present and the delayed positions of the tool in the form

$$h(t) = v_f \tau(t) + y(t - \tau(t)) - y(t) , \tag{5.7}$$

where v_f is the feed velocity. Thus, the governing equation can be written as

$$m\ddot{x}(t) + c_x \dot{x}(t) + k_x x(t) = K_x w \left(v_f \tau(t) + y(t - \tau(t)) - y(t) \right)^q , \tag{5.8}$$

$$m\ddot{y}(t) + c_y \dot{y}(t) + k_y y(t) = K_y w \left(v_f \tau(t) + y(t - \tau(t)) - y(t) \right)^q , \tag{5.9}$$

with $\tau(t + T) = \tau(t)$ defined by (5.6). It is assumed that there is a periodic solution $x_p(t+T) = x_p(t)$, $y_p(t+T) = y_p(t)$ that satisfies (5.8)–(5.9). The general solution can be written as $x(t) = x_p(t) + \xi(t)$ and $y(t) = y_p(t) + \eta(t)$ with $\xi(t)$ and $\eta(t)$ being perturbations around $x_p(t)$ and $y_p(t)$, respectively. Substitution into (5.8)–(5.9), expanding into power series with respect to $\xi(t)$ and $\eta(t)$, and eliminating the higher-order terms give the variational system in the form

$$m\ddot{\xi}(t) + c_x \dot{\xi}(t) + k_x \xi(t)$$
$$= K_x w q \left(v_f \tau(t) + y_p(t - \tau(t)) - y_p(t) \right)^{q-1} (\eta(t - \tau(t)) - \eta(t)) , \tag{5.10}$$

$$m\ddot{\eta}(t) + c_y \dot{\eta}(t) + k_y \eta(t)$$
$$= K_y w q \left(v_f \tau(t) + y_p(t - \tau(t)) - y_p(t) \right)^{q-1} (\eta(t - \tau(t)) - \eta(t)) . \tag{5.11}$$

Note that (5.10) is an ODE with state variable ξ forced by η, while (5.11) is a linear time-periodic DDE with state variable η. Since the homogeneous part of (5.10) is a simple damped oscillator, the stability of the system is determined by (5.11) only.

In the next step, we assume that the tool experiences only small oscillations such that the terms $y_p(t)$ and $y_p(t - \tau(t))$ can be neglected compared to the term $v_f \tau(t)$. Utilizing this assumption, the stability of the system is determined by the equation

$$m\ddot{\eta}(t) + c_y\dot{\eta}(t) + k_y\eta(t) = K_ywq(v_f\tau(t))^{q-1}\left(\eta(t - \tau(t)) - \eta(t)\right) . \qquad (5.12)$$

This is a DDE with a time-periodic coefficient and with time-periodic delay.

5.1.2 Constant Spindle Speed

In the case of constant spindle speed, (5.6) gives $\tau = 60/\Omega$, and the stability of the process is determined by the time-invariant DDE

$$\ddot{\eta}(t) + 2\zeta\omega_n\dot{\eta}(t) + \omega_n^2\eta(t) = H\left(\eta(t - \tau) - \eta(t)\right) , \qquad (5.13)$$

where $\omega_n = \sqrt{k_y/m}$ is the natural angular frequency, $\zeta = c_y/(2m\omega_n)$ is the damping ratio of the tool in the y direction, and $H = K_ywq(v_f\tau)^{q-1}/m$ is the specific cutting-force coefficient. Note that H is linearly proportional to the depth of cut w, which is an important technological parameter for the machinist. Equation (5.13) is the simplest mathematical model that describes regenerative machine tool chatter. In fact, it is a reparameterization of the delayed oscillator (2.28) in Section 2.2.

The stability of the system can be determined by the D-subdivision method. The characteristic equation reads

$$\lambda^2 + 2\zeta\omega_n\lambda + \omega_n^2 + H\left(1 - e^{-\lambda\tau}\right) = 0 . \qquad (5.14)$$

Substitution of $\lambda = \pm i\omega$ and decomposition into real and imaginary parts gives

$$\text{Re} : \quad -\omega^2 + \omega_n^2 + H - H\cos(\omega\tau) = 0 , \qquad (5.15)$$
$$\text{Im} : \quad 2\zeta\omega_n\omega + H\sin(\omega\tau) = 0 . \qquad (5.16)$$

Solving this system of equations for H and $\Omega = 60/\tau$ gives the D-curves in the parametric form

$$\Omega = \frac{30\omega}{j\pi - \arctan\left(\dfrac{\omega^2 - \omega_n^2}{2\zeta\omega_n\omega}\right)} , \quad j = 1, 2, \ldots , \qquad (5.17)$$

$$H = \frac{\left(\omega^2 - \omega_n^2\right)^2 + 4\zeta^2\omega_n^2\omega^2}{2\left(\omega^2 - \omega_n^2\right)} , \qquad (5.18)$$

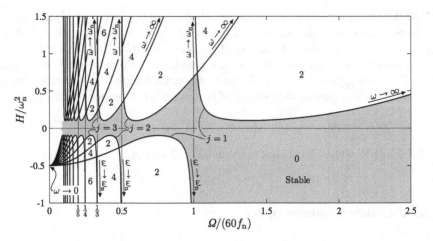

Fig. 5.3 Stability chart and the number of unstable characteristic exponents for (5.13) with $\zeta = 0.05$.

where the parameter ω is the frequency of the arising vibrations in [rad/s].

Figure 5.3 shows the D-curves and the number of unstable characteristic exponents in the plane of the dimensionless specific cutting-force coefficient

$$\frac{H}{\omega_n^2} = \frac{\left(\left(\frac{\omega}{\omega_n}\right)^2 - 1\right)^2 + 4\zeta^2 \left(\frac{\omega}{\omega_n}\right)^2}{2\left(\left(\frac{\omega}{\omega_n}\right)^2 - 1\right)} \tag{5.19}$$

and the dimensionless spindle speed $\Omega/(60f_n)$, where $f_n = \omega_n/2\pi$ is the natural frequency of the tool in [Hz]. As can be seen, (5.17) and (5.18) give a pair of D-curves for each integer j, one in the domain $H > 0$ associated with $\omega > \omega_n$ and one in the domain $H < 0$ associated with $\omega < \omega_n$. In the literature, these D-curves are called lobes or stability lobes. The lobes of index $j = 1$ are the rightmost ones; all the other lobes with $j \geq 2$ are located at lower spindle speeds. Each pair of lobes has a vertical asymptote at $\Omega = 60f_n/j$ indicated by dotted lines in Figure 5.3. The limits for the frequency parameter ω along the stability lobes are also presented. Frequency parameters $\omega < 0$ give D-curves in the negative spindle speed domain (not presented here). The number of unstable characteristic exponents can be determined by the analysis of the exponent-crossing direction along the D-curves in the same way as was shown in Section 2.2. The domain indicated by 0 in Figure 5.3 corresponds to asymptotic stability. Note that from a practical point of view, only the domains $H > 0$ are relevant, since these correspond to positive depth of cut values.

Figure 5.4 shows the practical region of the stability chart (i.e., for positive depths of cut) and the associated frequency ratio diagram, where $f = \omega/2\pi$ is the frequency of the arising self-excited vibrations at the stability boundaries in [Hz]. The stability lobes are usually characterized by their index: the lobe of index j is called the jth

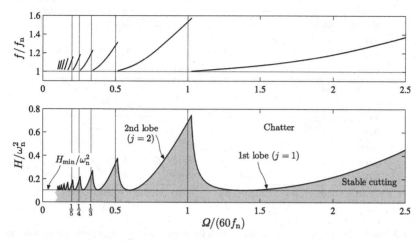

Fig. 5.4 Stability lobe diagram and frequency diagram for (5.13) with $\zeta = 0.05$.

lobe. The minimum points of the lobes can be determined by differentiating (5.18) with respect to ω, giving

$$\frac{dH}{d\omega}(\omega^*) = 0 \quad \Rightarrow \quad \omega^* = \omega_n \sqrt{2\zeta + 1}, \quad H_{\min} = H(\omega^*) = 2\zeta\omega_n^2(1 + \zeta). \quad (5.20)$$

For $\zeta = 0.05$ in Figure 5.4, (5.20) gives $H_{\min}/\omega_n^2 = 2\zeta(1 + \zeta) = 0.105$.

In the above model, machine tool chatter is related directly to the regenerative effect. It should be mentioned, however, that other types of self-excited vibrations may also occur in machining operations. For instance, the chip formation itself can produce oscillations in the cutting force for certain cutting speeds, as was shown by Burns and Davies [41, 42] and by Csernák and Pálmai [58, 218] for different models.

5.1.3 Varying Spindle Speed

Consider now the turning operation with periodically varying spindle speed. The process is described by (5.12) with the implicit equation (5.6) for the time-periodic regenerative delay $\tau(t)$. It is assumed that the spindle speed is modulated around an average value as

$$\Omega(t) = \Omega_0 + \Omega_1 S(t), \qquad S(t + T) = S(t), \quad (5.21)$$

where Ω_0 is the mean value, Ω_1 is the amplitude, and the time-periodic bounded function $S : \mathbb{R} \to [-1, 1]$ presents the shape of the variation. In general, (5.6) cannot be solved in closed form for τ. For instance, a cosine variation $S(t) = \cos(2\pi t/T)$

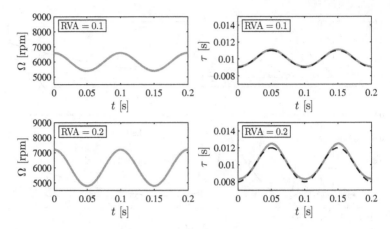

Fig. 5.5 Sinusoidal spindle speed variations with the corresponding exact (gray continuous line) and the approximate (black dashed line) time-delay variations for Ω_0 = 6000 rpm, RVF = 0.1 and RVA = 0.1 and 0.2.

gives the equation

$$\frac{1}{60}\left(\Omega_0\tau(t) + \frac{T}{2\pi}\Omega_1\left(\sin(2\pi t/T) - \sin\left(2\pi(t - \tau(t))/T\right)\right)\right) = 1 \ , \tag{5.22}$$

where $\tau(t)$ appears both explicitly and in the argument of a sine function, so it cannot be derived in closed form. Still, if Ω_1 is small enough compared to Ω_0, then the approximation

$$\tau(t) \approx \tau_0 - \tau_1 S(t) \tag{5.23}$$

can effectively be used, where $\tau_0 = 60/\Omega_0$ and $\tau_1/\tau_0 = \Omega_1/\Omega_0$. Figure 5.5 shows the exact and the approximated time delay for a harmonic spindle speed variation. Here, the notation RVA $:= \Omega_1/\Omega_0$ represents the ratio of the amplitude Ω_1 and the mean value Ω_0. In practical applications, the maximum value for RVA is about 0.2, which represents a variation of 20% of the spindle speed and results in a variation of 20% of the feed per revolution due to the constant feed velocity. Another parameter used in the literature is the ratio of the modulation frequency and the mean spindle speed, defined as RVF $:= 60/(\Omega_0 T)$. The maximum difference between the approximated and the exact time delays shown in Figure 5.5 is 1% for the case RVA = 0.1, and it is 4% for the case RVA = 0.2.

The stability of the system is described by (5.12), which can be simplified as

$$\ddot{\eta}(t) + 2\zeta\omega_n\dot{\eta}(t) + \omega_n^2\eta(t) = H_0\left(\frac{\tau(t)}{\tau_0}\right)^{q-1}\left(\eta(t - \tau(t)) - \eta(t)\right) \ , \tag{5.24}$$

where $\omega_n = \sqrt{k_y/m}$ is the natural angular frequency, $\zeta = c_y/(2m\omega_n)$ is the damping ratio, and $H_0 = K_y w q(v_f\tau_0)^{q-1}/m$ is a mean specific cutting force coefficient. The system can be written in the first-order form

$$\dot{x}(t) = A(t)x(t) + B(t)u(t - \tau(t)) , \qquad (5.25)$$
$$u(t) = Dx(t) , \qquad (5.26)$$

where

$$x(t) = \begin{pmatrix} \eta(t) \\ \dot{\eta}(t) \end{pmatrix} , \qquad D = \begin{pmatrix} 1 & 0 \end{pmatrix} , \qquad (5.27)$$

$$A(t) = \begin{pmatrix} 0 & 1 \\ -\left(\omega_n^2 + H_0 \left(\frac{\tau(t)}{\tau_0}\right)^{q-1}\right) & -2\zeta\omega_n \end{pmatrix} , \qquad B(t) = \begin{pmatrix} 0 \\ H_0 \left(\frac{\tau(t)}{\tau_0}\right)^{q-1} \end{pmatrix} . \qquad (5.28)$$

Stability charts for (5.25)–(5.26) can be determined by the semi-discretization method, as shown in Section 3.3. Figure 5.6 shows the stability chart in the plane of the dimensionless mean specific cutting-force coefficient H_0/ω_n^2 and the dimensionless spindle speed $\Omega/(60f_n)$ for the high spindle speed domains (for lobes of indices $j = 1, 2, 3, 4, 5$). The damping ratio is $\zeta = 0.02$. The diagrams were determined by the first-order semi-discretization method such that the delay resolution was $r = 44$ for all cases. The period resolutions corresponding to the frequency ratios RVF = $0.5, 0.2, 0.1$ and 0.05 were $p = 80, 200, 400$, and 800, respectively. (Note that the principal period of the system is $T = \tau_0/\text{RVF}$.) The resolution of the parameter plane $(\Omega/(60f_n), H_0/\omega_n^2)$ was 200×150. Since the monodromy matrix is obtained by $(p - 1)$ multiplications of $(r + 2) \times (r + 2)$ matrices at each point in the parameter plane, the computation of these stability charts is quite time-consuming. In these cases, the efficient matrix multiplication techniques presented in Sections 3.4.1 and 3.4.2 can be used effectively.

In Figure 5.6, the stability boundaries associated with constant-speed machining are also represented by dashed lines for reference. It can be seen that some improvements in the stability can be attained for RVF = 0.5, which corresponds to relatively fast spindle speed modulation. For realistic frequency ratios RVF = 0.2, 0.1, and 0.05, the difference between variable- and constant-speed machining is negligible in the high-speed domain.

Figure 5.7 shows the stability chart for the low spindle speed domains (for lobes of indices $j = 6, 7, 8, 9, 10$) with damping ratio $\zeta = 0.02$. Here, first-order semi-discretization was used with delay resolution $r = 110$. The period resolutions corresponding to the frequency ratios RVF = $0.5, 0.2, 0.1$, and 0.05 were $p = 200, 500, 1000$, and 2000, respectively. In this case, spindle speed variation has an advantageous effect on the stability of the process: the minimal depth of cut is about double that of constant-spindle-speed machining.

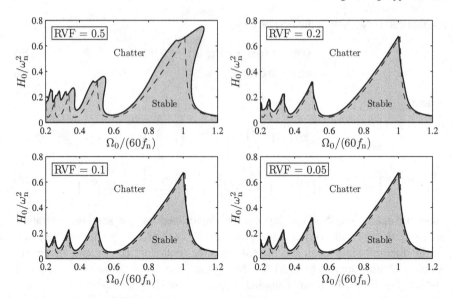

Fig. 5.6 Stability charts for turning process subjected to harmonic spindle speed modulation with RVA $= \Omega_1/\Omega_0 = 0.1$ in the high-speed domain. Dashed lines indicate the stability boundaries associated with constant-speed machining.

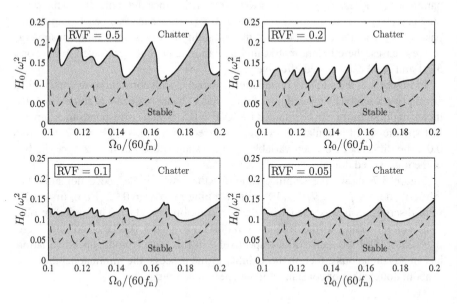

Fig. 5.7 Stability charts for turning process subjected to harmonic spindle speed modulation with RVA $= \Omega_1/\Omega_0 = 0.1$ in the low-speed domain. Dashed lines indicate the stability boundaries associated with constant-speed machining.

5.2 Stability of Milling Processes

Like turning processes, regenerative machine tool chatter is a challenging problem for milling operations. In the case of milling, surface regeneration is coupled with parametric excitation of the cutting teeth, resulting in a DDE with time-periodic coefficients. The first results regarding the stability properties of milling processes appeared in Tobias's and Tlusty's works [282, 281, 280]. They considered the time-averaged cutting force instead of the time-periodic one, and thus their models were equivalent to a DDE with constant coefficients. These models can be used for processes with large radial immersion and large number of cutting teeth when the parametric excitation of the teeth is negligible. For small radial immersion, however, the intermittent nature of the cutting process cannot be neglected. In these cases, time-domain simulations can be used to capture the dynamic behavior of the system for different technological parameters [140, 278, 253]. Minis and Yanushevsky [193] applied a Fourier expansion and Hill's infinite determinant technique in the frequency domain to derive stability charts for a two-degrees-of-freedom milling model. Budak and Altintas [6, 39, 40] used a similar approach for a multiple-degrees-of-freedom milling model. Their methods are often referred to as single-frequency (or zero-order) solution and multi-frequency solution, depending on the number of harmonics taken into account during the Fourier expansion. All these publications dealt with milling operations with large radial depth of cut and multiple cutting teeth such that the time-periodicity of the cutting force was not significant. In these models, the loss of stability is represented by a Hopf bifurcation similar to that of turning processes, i.e., a complex conjugate pair of characteristic exponents crosses the imaginary axis from left to right.

In the last decade, extended investigation of the high-speed milling process and the corresponding time-periodic DDE has led to the realization of a new bifurcation phenomenon. In addition to the stability lobes associated with Hopf bifurcation, new stability boundaries may appear representing a period-doubling (flip) bifurcation. Davies et al. [61, 62] modeled small radial immersion milling as an impact-like cutting process and obtained analytical formulas for the flip stability boundaries. Insperger and Stepan [121, 122] investigated a single-degree-of-freedom model of the milling process and demonstrated the appearance of flip stability lobes. They approximated the point delay in the model equations by a distributed delay with kernel function the gamma function. Zhao and Balachandran [305] determined stability charts by numerical simulations and showed period-doubling behavior at the stability boundaries. These results were confirmed by several other techniques: Bayly et al. [22, 23] used the temporal finite element method; Merdol and Altintas [186] used the multi-frequency solution; Szalai and Stepan [269, 270] determined the characteristic functions of the system and obtained stability criteria using the argument principle; Corpus and Endres [54, 55] reduced the problem to the flip boundaries where time-periodic ODEs describe the system instead of time-periodic DDEs; Butcher et al. [45] used an expansion of the solution in terms of Chebyshev polynomials. The semi-discretization method itself was developed in order to derive stability diagrams for the milling process [123, 129]. The existence of the period-doubling phenomenon was also confirmed by experiments in [61, 62, 23, 56, 180, 90].

5.2.1 Single-Degree-of-Freedom Model

One of the simplest models of end milling is shown in Figure 5.8. The workpiece is assumed to be flexible in the feed direction (direction x) with modal mass m, damping coefficient c, and spring stiffness k, while the tool is assumed to be rigid. The tool has N equally distributed cutting teeth with zero helix angles. The spindle speed is Ω given in [rpm]. According to Newton's law, the equation of motion reads

$$m\ddot{x}(t) + c\dot{x}(t) + kx(t) = -F_x(t) \,, \tag{5.29}$$

where $F_x(t)$ is the x component of the cutting force vector acting on the tool. Let the teeth of the tool be indexed by $j = 1, 2, \ldots, N$. The geometry of the milling process and the cutting forces are shown in Figure 5.9. The tangential and radial components of the cutting force acting on tooth j read

$$F_{j,\mathrm{t}}(t) = g_j(t)K_\mathrm{t}a_\mathrm{p}h_j^q(t) \,, \tag{5.30}$$

$$F_{j,\mathrm{r}}(t) = g_j(t)K_\mathrm{r}a_\mathrm{p}h_j^q(t) \,, \tag{5.31}$$

where K_t and K_r are the tangential and radial cutting-force coefficients, respectively, a_p is the axial depth of cut, $h_j(t)$ is the chip thickness cut by tooth j, and q is the cutting-force exponent. Function $g_j(t)$ is a screen function; it is equal to 1 if tooth j is in the cut, and 0 if it is not. If φ_{en} and φ_{ex} denote the angular locations where the cutting teeth enter and exit the cut, then the screen function reads

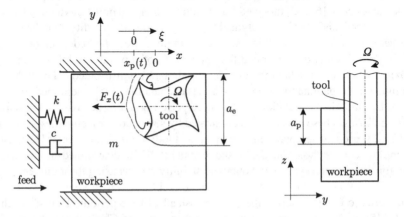

Fig. 5.8 Single-degree-of-freedom mechanical model of end milling process with a straight fluted tool.

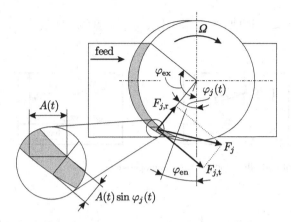

Fig. 5.9 Cutting-force components and chip thickness model in the milling process.

$$g_j(t) = \begin{cases} 1 & \text{if } \varphi_{\text{en}} < \varphi_j(t) \bmod 2\pi < \varphi_{\text{ex}}, \\ 0 & \text{otherwise}, \end{cases} \tag{5.32}$$

where

$$\varphi_j(t) = \frac{2\pi\,\Omega}{60}t + j\frac{2\pi}{N} \tag{5.33}$$

is the angular position of tooth j and mod is the modulo function. In the case of up-milling,

$$\varphi_{\text{en}} = 0, \qquad \varphi_{\text{ex}} = \arccos\left(1 - \frac{2a_e}{D}\right), \tag{5.34}$$

while in the case of down-milling,

$$\varphi_{\text{en}} = \arccos\left(\frac{2a_e}{D} - 1\right), \qquad \varphi_{\text{ex}} = \pi, \tag{5.35}$$

where a_e is the radial immersion and D is the diameter of the tool (see Figure 5.10).

Although a constant feed f_z per tooth is prescribed, the actual feed $A(t)$ per tooth is not constant, since it is affected by the present and a delayed position of the workpiece in the form

$$A(t) = f_z + x(t) - x(t - \tau), \tag{5.36}$$

where $\tau = 60/(N\Omega)$ [s] is the regenerative delay, which coincides with the tooth-passing period (note that the spindle speed Ω is given in [rpm]). The instantaneous chip thickness $h_j(t)$ is determined by the actual feed per tooth and the angular position of the cutting teeth. A circular approximation of the tooth path gives

$$h_j(t) = A(t)\sin\varphi_j(t) = (f_z + x(t) - x(t - \tau))\sin\varphi_j(t). \tag{5.37}$$

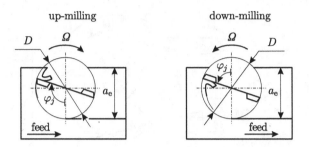

Fig. 5.10 Sketch of up-milling and down-milling operations.

Note that there exist more complex models for the chip thickness calculation, such as the trochoidal tooth path model, which results in time-dependent delays [80, 172] and models including the vibrations of the tool–workpiece system that results in state-dependent delays in the model equations [130, 17].

The x component of the cutting force acting on tooth j is obtained as the projection of $F_{j,t}$ and $F_{j,r}$ in the x direction, i.e.,

$$F_{j,x}(t) = F_{j,t}(t) \cos \varphi_j(t) + F_{j,r}(t) \sin \varphi_j(t) \ . \tag{5.38}$$

The x component of the resultant cutting force acting on the tool reads

$$F_x(t) = Q(t) (f_z + x(t) - x(t - \tau))^q \ , \tag{5.39}$$

where

$$Q(t) = \sum_{j=1}^{N} a_p g_j(t) \sin^q \varphi_j(t) \left(K_t \cos \varphi_j(t) + K_r \sin \varphi_j(t) \right) \tag{5.40}$$

is a periodic function with period τ. Thus, the equation of motion is the following nonlinear time-periodic DDE:

$$m\ddot{x}(t) + c\dot{x}(t) + kx(t) = -Q(t) (f_z + x(t) - x(t - \tau))^q \ . \tag{5.41}$$

Assume that the motion of the workpiece can be written in the form

$$x(t) = x_p(t) + \xi(t) \ , \tag{5.42}$$

where $x_p(t) = x_p(t + \tau)$ is a τ-periodic function, and $\xi(t)$ is the perturbation (see Figure 5.8). In fact, $x_p(t)$ is the solution for the unperturbed ideal case when no self-excited vibrations arise. Substitution of (5.42) into (5.41) yields

$$m\ddot{x}_p p(t) + c\dot{x}_p(t) + kx_p(t) + m\ddot{\xi}(t) + c\dot{\xi}(t) + k\xi(t)$$
$$= -Q(t) (f_z + \xi(t) - \xi(t - \tau))^q \ . \tag{5.43}$$

In the ideal unperturbed case, $\xi(t) \equiv 0$ and the oscillations of the workpiece are described by $x(t) = x_p(t)$. This case gives an ODE for x_p in the form

$$m\ddot{x}_p(t) + c\dot{x}_p(t) + kx_p(t) = -f_z^q Q(t) . \tag{5.44}$$

Since this is a linear differential equation with τ-periodic forcing, it has a τ-periodic particular solution. This proves the existence of the τ-periodic function $x_p(t)$ and verifies the decomposition (5.42).

For linear stability analysis, the variational system of (5.41) about the periodic motion $x_p(t)$ should be determined. Taylor expansion of the nonlinear terms in (5.43) with respect to ξ and neglecting the higher-order terms gives

$$m\ddot{x}_p(t) + c\dot{x}_p(t) + kx_p(t) + m\ddot{\xi}(t) + c\dot{\xi}(t) + k\xi(t)$$
$$= -f_z^q Q(t) - q f_z^{q-1} Q(t) \left(\xi(t) - \xi(t-\tau) \right) . \tag{5.45}$$

Using (5.44) and (5.45), a linear time-periodic DDE is obtained for ξ:

$$m\ddot{\xi}(t) + c\dot{\xi}(t) + k\xi(t) = -q f_z^{q-1} Q(t) \left(\xi(t) - \xi(t-\tau) \right) . \tag{5.46}$$

Introducing the natural angular frequency $\omega_n = \sqrt{k/m}$ and the damping ratio $\zeta = c/(2m\omega_n)$, (5.46) can be written in the form

$$\ddot{\xi}(t) + 2\zeta\omega_n\dot{\xi}(t) + \omega_n^2\xi(t) = -HG(t) \left(\xi(t) - \xi(t-\tau) \right) , \tag{5.47}$$

where $H = a_p q f_z^{q-1} K_r / m$ is the specific cutting-force coefficient and

$$G(t) = \sum_{j=1}^{N} g_j(t) \sin^q \varphi_j(t) \left(\frac{K_t}{K_r} \cos \varphi_j(t) + \sin \varphi_j(t) \right) \tag{5.48}$$

is a τ-periodic function called the directional dynamic cutting-force coefficient, or simply the directional factor. If $K_r = K_y$, $a_p = w$, $f_z = v_f\tau$, and $G(t) \equiv 1$, then (5.47) gives the governing equation (5.13) of constant-spindle-speed turning.

The system can be written in the first-order form

$$\dot{\mathbf{x}}(t) = \mathbf{A}(t)\mathbf{x}(t) + \mathbf{B}(t)\mathbf{u}(t - \tau) , \tag{5.49}$$
$$\mathbf{u}(t) = \mathbf{D}\mathbf{x}(t) , \tag{5.50}$$

where

$$\mathbf{x}(t) = \begin{pmatrix} \xi(t) \\ \dot{\xi}(t) \end{pmatrix} , \qquad \mathbf{D} = \begin{pmatrix} 1 & 0 \end{pmatrix} , \tag{5.51}$$

$$\mathbf{A}(t) = \begin{pmatrix} 0 & 1 \\ -\left(\omega_n^2 + HG(t) \right) & -2\zeta\omega_n \end{pmatrix} , \qquad \mathbf{B}(t) = \begin{pmatrix} 0 \\ HG(t) \end{pmatrix} . \tag{5.52}$$

Stability charts for (5.49)–(5.50) can be determined by the semi-discretization method, as shown in Section 3.3.

Figure 5.11 shows a series of stability lobe diagrams in the plane of the dimensionless spindle speed $N\Omega/(60f_n)$ and the dimensionless specific cutting-force coefficient H/ω_n^2 for different milling operations. (Note that H is linearly proportional to the axial depth of cut a_p.) The diagrams were determined by the first-order semi-discretization method with period resolution $p = 50$. The corresponding frequency diagrams and the directional factor $G(t)$ are also presented. The damping ratio is $\zeta = 0.02$, the cutting-force ratio is $K_t/K_r = 0.3$, and the cutting-force exponent is $q = 0.75$. The same diagrams are also shown for turning as the special limiting case when $G(t) \equiv 1$. The technological parameters for the milling operations were determined such that the time-dependency of the directional factor $G(t)$ becomes stronger and stronger. The first case is a full-immersion milling with a 4-fluted tool. In this case the tool is always in contact with the workpiece, since two of its cutting edges are always in the cut, and the directional factor $G(t)$ is a continuous function. The other cases are all up-milling operations by a 4-fluted tool with smaller and smaller radial immersion, resulting in more and more interrupted machining. As was shown by Davies et al. [61, 62], highly interrupted machining operations can be modeled approximately by a finite-dimensional discrete map instead of infinite-dimensional DDEs such that the cutting process is considered as an impact with the cutting force impulse being proportional to the chip thickness. In this sense, Figure 5.11 presents a transition between two special models of machining: the traditional time-independent DDE model of turning operation and the discrete map model of highly interrupted machining. Figure 5.11 shows that a series of extra stability lobes arises in addition to the Hopf lobes of turning as the process becomes more and more interrupted. Numerical calculation shows that along these additional lobes, a single characteristic multiplier crosses the unit circle at -1, i.e., these lobes are associated with period-doubling (flip) bifurcation. As the radial immersion decreases, the orientation of the flip lobes become vertical, and the stability diagrams tend to those of the highly interrupted model obtained by Davies et al. [61, 62].

As was mentioned in Chapter 1, the critical characteristic multipliers can be located in three ways:

1. $|\mu_{1,2}| = 1$ with $\operatorname{Im}\mu_{1,2} \neq 0$ (secondary Hopf bifurcation);
2. $\mu_1 = 1$ (cyclic-fold bifurcation);
3. and $\mu_1 = -1$ (period-doubling or flip bifurcation).

It can easily be seen that the case $\mu_1 = 1$ cannot occur for (5.47). It is known that in the critical subspace, $\xi(t + \tau) = \mu_1\xi(t)$ is satisfied. If $\mu_1 = 1$, then $\xi(t - \tau) = \xi(t)$, and substitution into (5.47) gives the damped oscillator

$$\ddot{\xi}(t) + 2\zeta\omega_n\dot{\xi}(t) + \omega_n^2\xi(t) = 0 . \tag{5.53}$$

Since ζ and ω_n are positive, (5.53) is asymptotically stable; consequently, it cannot have a characteristic exponent equal to zero, i.e., it cannot have a characteristic multiplier equal to 1. This proves that cyclic-fold bifurcation cannot arise for (5.47).

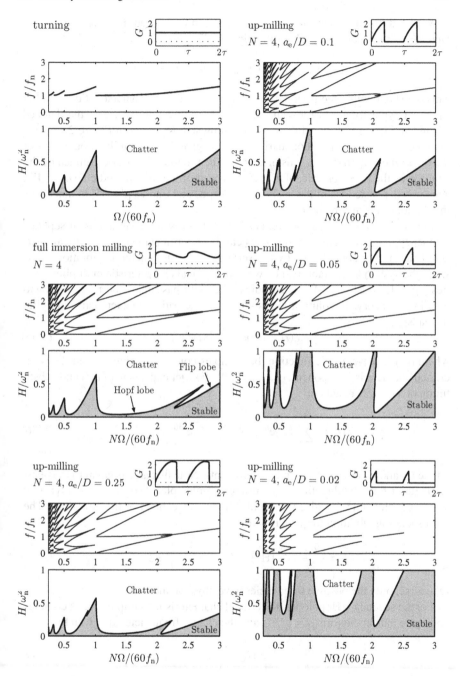

Fig. 5.11 Stability charts and frequency diagrams with the corresponding directional factor $G(t)$ for turning and different milling operations.

With a different conclusion, the same idea can be applied in the case $\mu_1 = -1$. Here, $\xi(t - \tau) = -\xi(t)$, and substitution into (5.47) gives

$$\ddot{\xi}(t) + 2\zeta\omega_n\dot{\xi}(t) + \left(\omega_n^2 + 2HG(t)\right)\xi(t) = 0 . \tag{5.54}$$

This is an ODE with time-periodic coefficient, for which the characteristic multiplier $\mu_1 = -1$ typically arises for some parameter combination (see, e.g., the damped Mathieu equation (2.64) and the corresponding stability chart in Figure 2.9). As a matter of fact, the stability boundaries of (5.54) give the flip stability boundaries of the original equation (5.47). This means that the flip lobes can be determined by the analysis of the time-periodic ODE (5.54) instead of the time-periodic DDE (5.47). This was the basic idea of the analysis by Corpus and Endres [54, 55] to determine the flip lobes for milling processes.

Figure 5.11 also shows the frequencies of the resulting vibrations in separate frequency diagrams. These frequencies can be determined using the critical characteristic multipliers obtained by the semi-discretization method. Vibrations arise when the system loses stability, i.e., when the critical characteristic multiplier satisfies $|\mu_1| = 1$. Equation (5.47) is periodic at the tooth-passing period τ. According to the Floquet theory, the solution corresponding to the critical characteristic multiplier μ_1 reads

$$\xi(t) = a(t)e^{\lambda_1 t} + \bar{a}(t)e^{\bar{\lambda}_1 t} , \tag{5.55}$$

where $a(t)$ is a τ-periodic function, bar denotes complex conjugate, and λ_1 is the critical characteristic exponent, i.e., $\mu_1 = e^{\lambda_1\tau}$. Fourier expansion of $a(t)$ and substitution of $\lambda_1 = i\omega_1$ results in

$$\xi(t) = \sum_{j=-\infty}^{\infty} \left(C_j e^{i(\omega_1 + j2\pi/\tau)t} + \bar{C}_j e^{-i(\omega_1 + j2\pi/\tau)t}\right) , \tag{5.56}$$

where C_j and $\bar{C}_j$ are some complex coefficients. Note that $\omega_1\tau$ is equal to the phase angle describing the direction of μ_1 in the complex plane, so that $-\pi < \omega_1\tau \leq \pi$. The exponents in (5.56) give the angular frequency content of the vibrations. The corresponding vibration frequencies are

$$f = \pm\frac{\omega_1}{2\pi} + \frac{j}{\tau} \text{ [Hz]} , \qquad j = 0, \pm1, \pm2, \ldots . \tag{5.57}$$

Of course, only the positive frequencies have physical meaning.

For the secondary Hopf lobes, the critical characteristic multipliers are a complex conjugate pair in the form $\mu = e^{\pm i\omega\tau}$, and the chatter frequencies are given by

$$f_H = \pm\frac{\omega_1}{2\pi} + j\frac{N\Omega}{60} \text{ [Hz]} , \qquad j = 0, \pm1, \pm2, \ldots . \tag{5.58}$$

According to (5.55), the solution is given as the product of the τ-periodic function $a(t)$ and the $(2\pi/\omega_1)$-periodic function $e^{\lambda_1 t} = e^{i\omega_1 t}$. Consequently, the resulting vibrations are quasiperiodic. In the literature, vibrations due to secondary Hopf bifurcations are often referred to as quasiperiodic chatter.

For the flip lobes, the critical characteristic multiplier is $\mu_1 = -1$, i.e., $\omega_1 \tau = \pi$. The corresponding chatter frequencies are

$$f_F = \frac{N\Omega}{120} + j\frac{N\Omega}{60} \text{ [Hz]}, \quad j = 0, \pm 1, \pm 2, \ldots . \tag{5.59}$$

In this case, the period of the function $e^{\lambda_1 t} = e^{i\omega_1 t} = e^{i\pi t/\tau}$ is 2τ. Consequently, the solution according to (5.55) is a 2τ-periodic function that explains the terminology period doubling: the period of the vibration is double the tooth-passing period.

The frequency diagrams in Figure 5.11 were obtained using (5.58) and (5.59). While turning operations are characterized by a well-defined single chatter frequency according to the Hopf bifurcation of autonomous systems, milling operations, being parametrically excited systems, present multiple vibration frequencies. Along the flip lobes, the basic frequency of the vibrations is equal to half of tooth-passing frequency. Along the Hopf lobes, quasiperiodic vibrations arise. It should be mentioned that flip instability is directly related to the time-periodic nature of the milling process. It occurs mostly for operations with small radial immersion when the directional factor $G(t)$ is strongly time-dependent.

Note that the frequency diagrams in Figure 5.11 do not distinguish the dominant vibration frequencies. Generally, only one or two of these frequencies characterize the chatter signal, and the other harmonics are associated with negligible amplitudes. A technique to show the strength of the different frequency components in complex milling models using the semi-discretization method was presented by Dombovari et al. [70].

Figure 5.12 shows a series of stability charts for different radial immersion ratios a_e/D for a 2-fluted end mill cutter. The damping ratio is $\zeta = 0.02$, the cutting-force ratio is $K_t/K_r = 0.3$, and the cutting-force exponent is $q = 0.75$. The plots give a transition of the stability charts between up-milling and down-milling. For up-milling operations, the Hopf lobes are located to the left of the flip lobes. As the radial immersion is increased, the flip lobes open and new Hopf stability boundaries appear while the original Hopf lobe shrinks to a loop-like curve (see the full-immersion case). Reducing the radial immersion at the other side of the workpiece leads to the down-milling operations (see Figure 5.10). As the radial immersion is decreased, the loop-like lobe disappears, while the new Hopf lobes remain dominant. For down-milling operations, the Hopf lobes are located to the right of the flip lobes.

Figure 5.12 clearly shows the main differences between up- and down-milling operations with a 2-fluted cutter. As can be seen, the flip lobes are located more or less at the same spindle speed ranges, although they may vary in size for the different cases. This is not true for the Hopf lobes. For low-immersion up-milling operations, the Hopf lobes are located to the left of the flip lobes, while for down-milling, the Hopf lobes are positioned to the right of the flip lobes. The physical explanation for this special duality is the following. The flip lobes are related to the impact effects of the cutting teeth as they enter and leave the cut. These are more or less

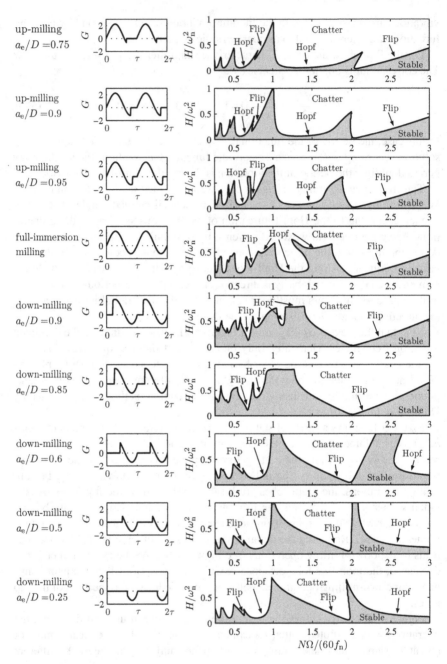

Fig. 5.12 Stability charts and the corresponding directional factor $G(t)$ for milling operations with different radial immersions for a 2-fluted tool ($N = 2$).

independent of the sense (up or down) of the operation. This is not the case for the Hopf lobes. As was shown in Figure 5.3, the conventional stability chart of turning operations contains a part for negative depths of cut, which has no physical meaning there. In the case of milling, the cutting force is multiplied by the directional factor $G(t)$, which is mostly positive for up-milling and mostly negative for down-milling operations (see Figure 5.12). The new Hopf lobes that emerge as the operation turns from up-milling to down-milling are related to the lobes in the negative depth of cut values in Figure 5.3, which are dual to those in the positive region. This is the explanation for the duality in the stability properties of up- and down-milling.

In addition to the intricate transition between the up- and down-milling cases, the stability chart of (5.47) shows another interesting feature: the flip stability boundaries are not always open lobes, but for certain parameter combinations, they are closed curves forming unstable islands in the stability charts. This phenomenon was shown by Szalai and Stepan [269, 270] for highly interrupted cutting. They showed in a precise analytic way that all the flip stability boundaries are in fact lense-shaped closed curves, except the first lobe at the highest cutting speeds. When these closed curves are intersected by other Hopf boundaries, they then appear as open stability lobes. If the ratio of time spent cutting to delay period tends to zero, then their model gives the results of Davies et al. [61, 62], as shown in [271]. Figure 5.13 shows the stability charts for an up-milling operation with radial immersion $a_e/D = 0.02$ by a 4-fluted tool for different damping-ratio parameters. It can be seen that as the damping ratio gets larger, the stability domains grow and the second flip lobe at spindle

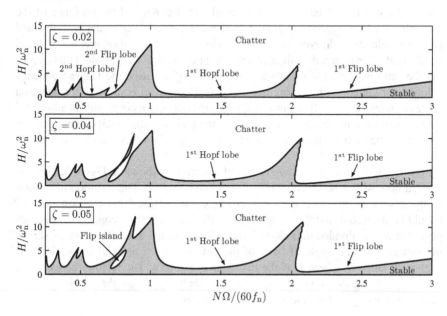

Fig. 5.13 Stability charts for an up-milling operation with radial immersion $a_e/D = 0.02$ by a 4-fluted tool for different damping ratio parameters ζ.

speed $N\Omega/(60f_n) \approx 0.7$ closes and forms an unstable island within the stable domain. These unstable islands appear as a result of the parametric forcing of the milling process, and are therefore referred to as *parametrically induced flip islands*.

5.2.2 Milling with Helical Tools

In the case of helical tools, the forces acting on the cutting edges change along the axial direction of the tool. In the case of large axial immersion, this effect should also be taken into account. The cutting force characteristics for helical tools are well known in the literature (see, e.g., [279, 5, 78, 231]), but results regarding the corresponding stability lobe diagrams have appeared only in recent years. Stability charts for a detailed model of end milling processes with helical tools were determined by Balachandran and Zhao [19, 305] by means of time-domain simulations. However, they analyzed low axial depths of cut, where the effect of the helical edges is negligible. A direct comparison of helical tools and straight fluted tools for large axial immersion were performed by Zatarain et al. [303]. They showed that there is a qualitative difference between the stability charts of helical tools and those of straight fluted tools: in the case of helical tools, the flip lobes may turn to closed curves forming unstable islands in the stability charts. They used both the multi-frequency solution [39, 186] and the semi-discretization method to construct stability charts and verified the results by experiments. Since then, several results have been presented for helical tools. Mann et al. [181] confirmed the presence of the unstable islands using the temporal finite element method. Wan et al. [293] obtained similar results for a different model using the semi-discretization method. Sims et al. [249, 301] analyzed the effect of varying pitch angles and varying helix angles via the semi-discretization method. Turner et al. [285] used the single-frequency solution for cutters with varying helix angles. Dombovari et al. [69] considered a more complex case, a helical cutter with serrated edges that contribute an additional waviness of the cutting edges in the radial direction (see also [185]). They used the semi-discretization method to construct stability charts for this complex tool geometry.

The single-degree-of-freedom model of end milling shown in Figure 5.8 is considered now with a helical tool of uniform helix angle β. The equation of motion of the system is the same as (5.29), but here, the cutting-force component $F_x(t)$ should be derived in a different way due to the helical cutting edges. For this purpose, the tool is divided into elementary disks along the axial direction, as shown in Figure 5.14. The angle of twist of the cutting edge j at axial immersion z is $\psi(z) = 2z \tan\beta/D$, where D is the diameter of the tool. The relation between the helix angle β and the helix pitch l_p is $\tan\beta = D\pi/(Nl_p)$. Thus, the angular position of the cutting edges along the axial direction reads

$$\varphi_j(t, z) = \frac{2\pi\Omega}{60}t + j\frac{2\pi}{N} - z\frac{2\pi}{Nl_p} \, . \tag{5.60}$$

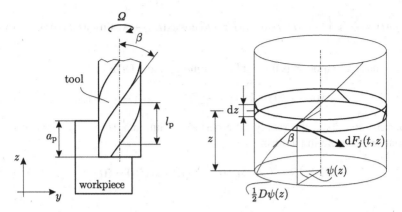

Fig. 5.14 Geometry of a helical tool and its division into elementary disks along the axial z direction.

The elementary cutting-force components acting on tooth j at a disk element of width dz are

$$dF_{j,t}(t,z) = g_j(t,z)K_t h^q(t,z)dz , \tag{5.61}$$

$$dF_{j,r}(t,z) = g_j(t,z)K_r h^q(t,z)dz , \tag{5.62}$$

where K_t and K_r are the tangential and radial cutting-force coefficients, respectively, $h_j(t,z)$ is the chip thickness cut by tooth j at axial immersion z, and q is the cutting-force exponent. Function $g_j(t,z)$ reads

$$g_j(t,z) = \begin{cases} 1 & \text{if } \varphi_{en} < \varphi_j(t,z) \bmod 2\pi < \varphi_{ex} , \\ 0 & \text{otherwise} , \end{cases} \tag{5.63}$$

where φ_{en} and φ_{ex} are the entrance and exit immersion angles. The instantaneous chip thickness at axial immersion z can be given as

$$h_j(t,z) = (f_z + x(t) - x(t - \tau)) \sin \varphi_j(t,z) , \tag{5.64}$$

where f_z is the feed per tooth and $\tau = 60/(N\Omega)$ [s] is the tooth-passing period. The x component of the elementary cutting force acting on tooth j reads

$$dF_{j,x}(t,z) = dF_{j,t}(t,z) \cos \varphi_j(t,z) + dF_{j,r}(t,z) \sin \varphi_j(t,z) , \tag{5.65}$$

and the x component of the resultant cutting force acting on the tool is

$$F_x(t) = Q(t) (f_z + (x(t) - x(t - \tau)))^q , \tag{5.66}$$

where

$$Q(t) = \sum_{j=1}^{N} \left(\int_0^{a_p} g_j(t,z) \sin^q \varphi_j(t,z) \Big(K_t \cos \varphi_j(t,z) + K_r \sin \varphi_j(t,z) \Big) dz \right) . \quad (5.67)$$

The equation of motion is the nonlinear time-periodic DDE

$$m\ddot{x}(t) + c\dot{x}(t) + kx(t) = -Q(t)\,(f_z + x(t) - x(t - \tau))^q . \quad (5.68)$$

Linearization about the steady-state periodic solution $x_p(t) = x_p(t + \tau)$ gives the variational system in the form

$$\ddot{\xi}(t) + 2\zeta\omega_n\dot{\xi}(t) + \omega_n^2\xi(t) = -G(t,a_p)\,(\xi(t) - \xi(t - \tau)) , \quad (5.69)$$

where

$$G(t,a_p) = \sum_{j=1}^{N} \left(\frac{qf_z^{q-1}}{m} \int_0^{a_p} g_j(t,z) \sin^q \varphi_j(t,z) \Big(K_t \cos \varphi_j(t,z) + K_r \sin \varphi_j(t,z) \Big) dz \right)$$

$$(5.70)$$

is a particular directional factor that depends on the axial immersion a_p. Equation (5.69) is written in the first-order form

$$\dot{x}(t) = A(t)x(t) + B(t)u(t - \tau) , \quad (5.71)$$

$$u(t) = Dx(t) , \quad (5.72)$$

with

$$x(t) = \begin{pmatrix} \xi(t) \\ \dot{\xi}(t) \end{pmatrix} , \qquad D = \begin{pmatrix} 1 & 0 \end{pmatrix} , \quad (5.73)$$

$$A(t) = \begin{pmatrix} 0 & 1 \\ -\big(\omega_n^2 + G(t,a_p)\big) & -2\zeta\omega_n \end{pmatrix} , \qquad B(t) = \begin{pmatrix} 0 \\ G(t,a_p) \end{pmatrix} . \quad (5.74)$$

Stability charts can be determined by the semi-discretization method, as shown in Section 3.3.

Stability charts are presented in Figure 5.15 for up-milling operations with different helix pitches. A 4-fluted tool is considered ($N = 4$) with diameter $D = 20\,\text{mm}$. The radial immersion is $a_e = 2\,\text{mm}$; thus, the radial immersion ratio is $a_e/D = 0.1$. The cutting-force coefficients are $K_t = 107 \times 10^6\,\text{N/m}^{1+q}$ and $K_r = 40 \times 10^6\,\text{N/m}^{1+q}$; the cutting-force exponent is $q = 0.75$. The feed per tooth is $f_z = 0.1\,\text{mm}$, for which the linearized cutting-force coefficients are $K_t q f_z^{q-1} = 800 \times 10^6\,\text{N/m}^2$ and $K_r q f_z^{q-1} = 300 \times 10^6\,\text{N/m}^2$. The stiffness is $k = 20 \times 10^6\,\text{N/m}$, the natural frequency is $f_n = \omega_n/2\pi = 400\,\text{Hz}$, and the damping ratio is $\zeta = 0.02$. The first panel shows the stability diagram for a straight fluted tool. This diagram corresponds to the case $a_e/D = 0.1$ in Figure 5.11. The other panels in Figure 5.15 show the stability charts for milling tools with different helix pitches. It can be seen that the flip lobes turn to closed curves forming unstable islands within the stable domain. These islands are

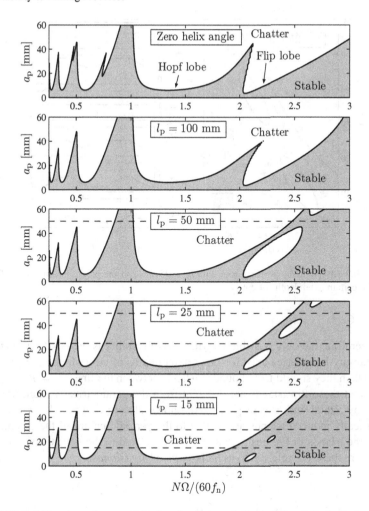

Fig. 5.15 Stability charts for a straight fluted tool (zero helix angle) and for helical tools with different helix pitches.

separated by the lines where the depth of cut is equal to the multiples of the helix pitch l_p (indicated by dashed lines in Figure 5.15).

The analysis of the particular directional factor $G(t, a_p)$ gives a clear explanation for the existence of these unstable islands. If the axial depth of cut is equal to a multiple of the helix pitch, that is, $a_p = jl_p$ with j being a positive integer, then $G(t, a_p)$ becomes constant in time, since the variation of the cutting forces distributed along the helical edges are balanced along the helix pitch l_p. Figure 5.16 shows the unstable islands for $l_p = 25$ mm and the plots of the particular directional factor $G(t, a_p)$ for different depths of cut a_p. For $a_p = l_p$ and $a_p = 2l_p$, $G(t, a_p)$ is constant in time. Consequently, at these parameters, the system is described by an autonomous

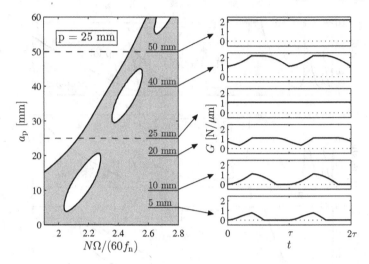

Fig. 5.16 Stability chart for a helical tool with $l_p = 25$ mm and the particular directional factor $G(t, a_p)$ for certain axial depths of cut a_p.

DDE. Therefore, the only possible way for loss of stability is the Hopf bifurcation. If flip stability boundaries exist for other axial depths of cut (when $a_p \neq jl_p$, $j = 1, 2, \ldots$) then they should form bounded islands between the lines $a_p = jl_p$.

Note that these flip islands are different from the parametrically induced ones reported by Szalai and Stepan [269, 270] shown also in Figure 5.13. While the parametrically induced islands are related to the periodic nature of the machining process, the islands in Figures 5.15 and 5.16 are related to the helical edges of the tool, and are therefore referred to as *helix-induced flip islands*.

5.2.3 Varying Spindle Speed Milling

Varying spindle speeds are also applied in milling operations in order to avoid chatter and thus to increase productivity. The idea behind this is that each flute experiences a different regenerative delay. In this way, the regenerative effect is disturbed, which may reduce the self-excited vibrations for certain spindle speeds. (A similar technique to disturb the regenerative effect is the application of variable pitch angles [7, 37, 38, 249, 293, 237], or serrated tools [185, 69].)

Mathematical models for milling processes with spindle speed variation are more complex than those of turning operations, since the speed-variation frequency and tooth-passing frequency interact and the resulting system is typically quasiperiodic. However, if the ratio of the mean period of the spindle speed and the speed modulation period is a rational number, then the system is purely time-periodic, and the Floquet theory can be applied to determine stability properties. Sastry et al. [229]

used Fourier expansion and applied the Floquet theory to derive stability lobe diagrams for face milling. They obtained some improvements for low spindle speeds. Zatarain et al. [304] presented a general method in the frequency domain for the problem, and showed that variable spindle speed can effectively be used for chatter suppression for low cutting speeds. They validated their model using the semi-discretization method and time-domain simulations, and confirmed their results also by experiments. Seguy et al. [236] used the semi-discretization method to analyze the stability properties around the first flip lobe.

Here, the mechanical model in Figure 5.8 is analyzed with time-periodic spindle speed modulation in the form

$$\Omega(t) = \Omega_0 + \Omega_1 S(t) , \qquad S(t + T) = S(t) , \tag{5.75}$$

where Ω_0 is the mean value, Ω_1 is the amplitude, and the time-periodic bounded function $S : \mathbb{R} \to [-1, 1]$ presents the shape of the variation. It is assumed that $\Omega(t + T) = \Omega(t) > 0$. The variation of the regenerative delay $\tau(t)$ can approximately be given in the implicit form

$$\int_{t-\tau(t)}^{t} \frac{\Omega(s)}{60} \, ds = \frac{1}{N} , \tag{5.76}$$

where N is the number of the cutting teeth. Similarly to turning operations (see (5.6), (5.22), and Figure 5.5), the time delay $\tau(t)$ cannot be expressed in closed form, but it can be approximated by

$$\tau(t) \approx \tau_0 - \tau_1 S(t) \tag{5.77}$$

with $\tau_0 = 60/(N\Omega_0)$ and $\tau_1/\tau_0 = \Omega_1/\Omega_0$ if Ω_1 is small enough compared to Ω_0.

The linearized equation of motion reads

$$\ddot{\xi}(t) + 2\zeta\omega_n\dot{\xi}(t) + \omega_n^2\xi(t) = -G(t)\,(\xi(t) - \xi(t - \tau(t))) , \tag{5.78}$$

where $\omega_n = \sqrt{k/m}$ and $\zeta = c/(2m\omega_n)$ are the natural angular frequency and the damping ratio, respectively,

$$G(t) = a_p \frac{q(v_f\tau(t))^{q-1}}{m} \sum_{j=1}^{N} g_j(t) \sin^q \varphi_j(t) \left(K_t \cos \varphi_j(t) + K_r \sin \varphi_j(t)\right) \tag{5.79}$$

is the specific directional factor, and v_f is the feed velocity. The screen function $g_j(t)$ is defined by (5.32) and the angular position of tooth j is

$$\varphi_j(t) = \frac{2\pi\,\Omega(t)}{60}t + j\,\frac{2\pi}{N} . \tag{5.80}$$

We assume that the ratio of the modulation period T and the mean time delay $\tau_0 = 60/(N\Omega_0)$ is a rational number, i.e., $q_1 T = q_2 \tau_0$ with q_1 and q_2 being relatively prime. In this case, the specific directional factor $G(t)$ is periodic with period $q_1 T$; consequently, the principal period of the system is also $q_1 T$. If the ratio T/τ_0 is not rational, then the system is quasiperiodic and the Floquet theory does not apply. The system is written in the first-order form

$$\dot{\mathbf{x}}(t) = \mathbf{A}(t)\mathbf{x}(t) + \mathbf{B}(t)\mathbf{u}(t - \tau(t)) \,, \tag{5.81}$$

$$\mathbf{u}(t) = \mathbf{D}\mathbf{x}(t) \,, \tag{5.82}$$

with

$$\mathbf{x}(t) = \begin{pmatrix} \xi(t) \\ \dot{\xi}(t) \end{pmatrix} \,, \qquad \mathbf{D} = \begin{pmatrix} 1 & 0 \end{pmatrix} \,, \tag{5.83}$$

$$\mathbf{A}(t) = \begin{pmatrix} 0 & 1 \\ -\left(\omega_n^2 + G(t)\right) & -2\zeta\omega_n \end{pmatrix} \,, \qquad \mathbf{B}(t) = \begin{pmatrix} 0 \\ G(t) \end{pmatrix} \,. \tag{5.84}$$

Stability charts can be determined by the semi-discretization method, as shown in Section 3.3.

The variation of the spindle speed is described by the amplitude ratio RVA = Ω_1/Ω_0 and the frequency ratio RVF = $60/(\Omega_0 T)$. Figure 5.17 shows some sample stability charts for RVA = 0.1 and for different RVF values in the high-speed domain. The parameters are as follows. A half-immersion up-milling process is considered (i.e., $a_e/D = 0.5$) by a 4-fluted tool ($N = 4$) with zero helix angle. The cutting-force coefficients are $K_t = 107 \times 10^6 \, \text{N/m}^{1+q}$ and $K_r = 40 \times 10^6 \, \text{N/m}^{1+q}$, the cutting-force exponent is $q = 0.75$. The feed per tooth is $f_z = 0.1 \, \text{mm}$, for which the linearized cutting-force coefficients are $K_t q f_z^{q-1} = 800 \times 10^6 \, \text{N/m}^2$ and $K_r q f_z^{q-1} = 300 \times 10^6 \, \text{N/m}^2$. The stiffness is $k = 20 \times 10^6 \, \text{N/m}$, the natural frequency is $f_n = \omega_n/2\pi = 400 \, \text{Hz}$, and the damping ratio is $\zeta = 0.02$. Using the assumption $q_1 T = q_2 \tau_0$ with q_1 and q_2 being relatively prime, one obtains $q_1/q_2 = \text{RVF}/N$. For the frequency ratios RVF = $0.5, 0.2, 0.1$, and 0.05 in Figure 5.17, these relatively prime numbers are $q_1 = 1$ and $q_2 = 8, 20, 40$, and 80, respectively. The corresponding principal periods are $q_1 T = 8\tau_0, 20\tau_0, 40\tau_0$, and $80\tau_0$. The diagrams were determined by the first-order semi-discretization method such that the delay resolution was $r = 44$ for all cases. The period resolutions corresponding to the frequency ratios RVF = $0.5, 0.2, 0.1$, and 0.05 were $p = 320, 800, 1600$, and 3200, respectively. (Note that the principal period of the system is $q_1 T = q_1 N \tau_0/\text{RVF}$.) Since the monodromy matrix is obtained by $(p-1)$ multiplications of $(r+2) \times (r+2)$ matrices, the efficient matrix multiplication techniques presented in Sections 3.4.1 and 3.4.2 have an important role here in saving computational time.

In Figure 5.17, dashed lines indicate the stability boundaries corresponding to constant-spindle-speed machining. As can be seen, spindle speed modulation results in some improvements in the stability around the first flip lobe.

Figure 5.18 shows the same stability charts for lower spindle speeds. Here, the first-order semi-discretization method was used with delay resolution $r = 110$. The period resolutions corresponding to the frequency ratios RVF = $0.5, 0.2, 0.1$, and 0.05 were $p = 800, 2000, 4000$, and 8000, respectively. In this case, spindle speed modulation results in an increase of the stability lobes in general. For small RVF values (i.e., for small modulation frequencies), the peaks of the stability boundaries

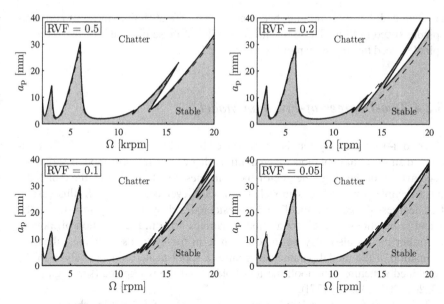

Fig. 5.17 Stability charts for milling processes with sinusoidal spindle speed modulation with RVA = 0.1 in the high-speed domain. Dashed lines indicate the stability boundaries associated with constant-spindle-speed milling.

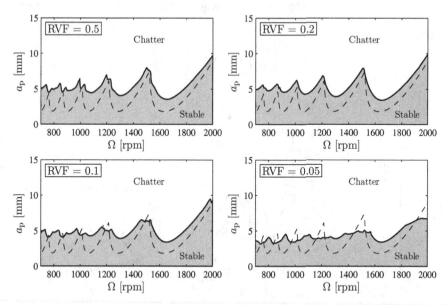

Fig. 5.18 Stability charts for milling processes with sinusoidal spindle speed modulation with RVA = 0.1 in the low-speed domain. Dashed lines indicate the stability boundaries associated with constant-spindle-speed milling.

are flattened, but their minima become larger by 50–60% than those of constant-spindle-speed milling. These results are similar to those obtained for the variable-spindle-speed turning shown in Figures 5.6 and 5.7.

5.2.4 Two-Degrees-of-Freedom Model

If the tool–workpiece system is characterized by multiple modes instead of a well-defined single dominant mode, then multiple-degrees-of-freedom models should be used instead of the single-degree-of-freedom ones. In a typical case, the tool is the most flexible part, and it is modeled as a cantilever beam [151]. In this case, a two-degrees-of-freedom model can be considered with symmetric parameters in the x and y directions, and diagonal modal matrices arise in the equation of motion. If further modes also play an important role in the system's dynamics, then the estimation of the modal parameters requires a sophisticated modal analysis of the combined structure—the tool, the tool-holder, and the workpiece (see, e.g., [232, 33, 201, 202, 235, 69, 177]).

Here, a two-degrees-of-freedom mechanical model of end milling shown in Figure 5.19 is analyzed. The tool is assumed to be flexible relative to the rigid workpiece. The governing equation has the form

$$\mathbf{M\ddot{q}}(t) + \mathbf{C\dot{q}}(t) + \mathbf{Kq}(t) = \mathbf{F}(t) , \qquad (5.85)$$

where

$$\mathbf{q}(t) = \begin{pmatrix} x(t) \\ y(t) \end{pmatrix} , \quad \mathbf{F}(t) = \begin{pmatrix} F_x(t) \\ F_y(t) \end{pmatrix} ,$$

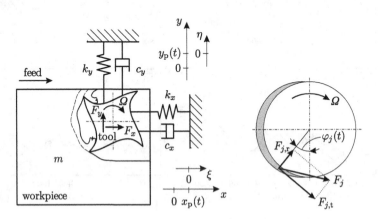

Fig. 5.19 Mechanical model of the tool–workpiece system in case of flexible tool (left) and the orientation of the cutting-force components acting on tooth j (right).

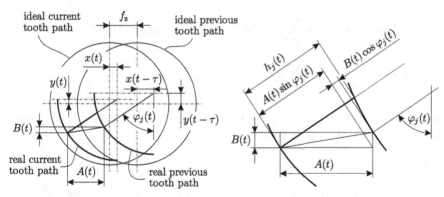

Fig. 5.20 Chip thickness calculation for two-degrees-of-freedom milling model.

are the position and the cutting force vectors, and **M**, **C** and **K** are the modal mass, damping, and stiffness matrices, respectively. If the tool has two uncoupled perpendicular modes, like a symmetric circular beam, then the modal matrices are diagonal with the same diagonal values. The x and y components of the cutting force on tooth j are given as

$$F_{j,x}(t) = F_{j,t}(t) \cos \varphi_j(t) + F_{j,r}(t) \sin \varphi_j(t) , \qquad (5.86)$$

$$F_{j,y}(t) = -F_{jt}(t) \sin \varphi_j(t) + F_{j,r}(t) \cos \varphi_j(t) , \qquad (5.87)$$

where $F_{j,t}(t)$ and $F_{j,r}(t)$ are the tangential and radial cutting-force components given in (5.30) and (5.31), respectively, and $\varphi_j(t)$ is the angular position of tooth j according to (5.33).

Since the tool experiences vibrations in both the feed and perpendicular directions, the instantaneous chip thickness is affected by the present and a delayed position of the tool in both the x and y directions, as shown in Figure 5.20. Assuming a circular approximation of the tooth path, the chip thickness can be given as

$$h_j(t) \approx A(t) \sin \varphi_j(t) + B(t) \cos \varphi_j(t)$$
$$= (f_z + x(t-\tau) - x(t)) \sin \varphi_j(t) + (y(t-\tau) - y(t)) \cos \varphi_j(t) , \quad (5.88)$$

where $A(t) = f_z + x(t-\tau) - x(t)$ is the actual feed per tooth, and $B(t) = y(t-\tau) - y(t)$. In Figure 5.20, the ideal tooth path is associated with the motion of the teeth of an ideally rigid tool, while the actual tooth path is associated with the motion of the teeth of the real flexible tool. The components of the resultant cutting force acting on the tool are

$$F_x(t) = \sum_{j=1}^{N} F_{j,x}(t) = \sum_{j=1}^{N} g_j(t) \left(K_t \cos \varphi_j(t) + K_r \sin \varphi_j(t) \right) a_p h_j^q(t) , \qquad (5.89)$$

$$F_y(t) = \sum_{j=1}^{N} F_{j,y}(t) = \sum_{j=1}^{N} g_j(t) \left(-K_t \sin \varphi_j(t) + K_r \cos \varphi_j(t) \right) a_p h_j^q(t) \qquad (5.90)$$

with $g_j(t)$ given in (5.32). Using (5.89), (5.90) and (5.88), the cutting force vector in (5.85) can be written in the form

$$\mathbf{F}(t) = \mathbf{F}(t, \mathbf{q}(t), \mathbf{q}(t - \tau))$$
$$= \begin{pmatrix} \sum_{j=1}^{N} S_{j,x}(t) \left((f_z + x(t - \tau) - x(t)) \sin \varphi_j(t) + (y(t - \tau) - y(t)) \cos \varphi_j(t) \right)^q \\ \sum_{j=1}^{N} S_{j,y}(t) \left((f_z + x(t - \tau) - x(t)) \sin \varphi_j(t) + (y(t - \tau) - y(t)) \cos \varphi_j(t) \right)^q \end{pmatrix} ,$$
$$(5.91)$$

where

$$S_{j,x}(t) = a_p g_j(t) \left(K_t \cos \varphi_j(t) + K_r \sin \varphi_j(t) \right) , \qquad (5.92)$$

$$S_{j,y}(t) = a_p g_j(t) \left(-K_t \sin \varphi_j(t) + K_r \cos \varphi_j(t) \right) . \qquad (5.93)$$

Equations (5.85) and (5.91) form a nonlinear time-periodic DDE. Similarly to the single-degree-of-freedom model, the solution can be decomposed as

$$\mathbf{q}(t) = \mathbf{q}_p(t) + \boldsymbol{\epsilon}(t) , \qquad (5.94)$$

where

$$\mathbf{q}_p(t) = \begin{pmatrix} x_p(t) \\ y_p(t) \end{pmatrix} \qquad (5.95)$$

is a τ-periodic forced component of the tool motion and

$$\boldsymbol{\epsilon}(t) = \begin{pmatrix} \xi(t) \\ \eta(t) \end{pmatrix} \qquad (5.96)$$

is the perturbation around $\mathbf{q}_p(t)$ (see Figure 5.19). Substitution of (5.94) into (5.85) and (5.91) and Taylor expansion with respect to $\boldsymbol{\epsilon}$ neglecting the higher-order terms gives the variational system in the form

$$\mathbf{M}\ddot{\boldsymbol{\epsilon}}(t) + \mathbf{C}\dot{\boldsymbol{\epsilon}}(t) + \mathbf{K}\boldsymbol{\epsilon}(t) = \mathbf{G}(t) \left(\boldsymbol{\epsilon}(t - \tau) - \boldsymbol{\epsilon}(t) \right) , \qquad (5.97)$$

where

$$\mathbf{G}(t) = \begin{pmatrix} G_{xx}(t) & G_{xy}(t) \\ G_{yx}(t) & G_{yy}(t) \end{pmatrix} \qquad (5.98)$$

is the specific directional factor matrix with the elements

$$G_{xx}(t) = a_p q f_z^{q-1} \sum_{j=1}^{N} g_j(t) \left(K_t \cos \varphi_j(t) + K_r \sin \varphi_j(t) \right) \sin^q \varphi_j(t) , \tag{5.99}$$

$$G_{xy}(t) = a_p q f_z^{q-1} \sum_{j=1}^{N} g_j(t) \left(K_t \cos \varphi_j(t) + K_r \sin \varphi_j(t) \right) \cos \varphi_j(t) \sin^{q-1} \varphi_j(t) ,$$

$$\tag{5.100}$$

$$G_{yx}(t) = a_p q f_z^{q-1} \sum_{j=1}^{N} g_j(t) \left(-K_t \sin \varphi_j(t) + K_r \cos \varphi_j(t) \right) \sin^q \varphi_j(t) , \tag{5.101}$$

$$G_{yy}(t) = a_p q f_z^{q-1} \sum_{j=1}^{N} g_j(t) \left(-K_t \sin \varphi_j(t) + K_r \cos \varphi_j(t) \right) \cos \varphi_j(t) \sin^{q-1} \varphi_j(t) .$$

$$\tag{5.102}$$

The stability of the periodic motion $\mathbf{q}_p(t)$, i.e., the stability of the machining process, is determined by (5.97). The periodic component $\mathbf{q}_p(t)$ of the tool motion can be obtained as the steady-state solution of

$$\mathbf{M}\ddot{\mathbf{q}}_p(t) + \mathbf{C}\dot{\mathbf{q}}_p(t) + \mathbf{K}\mathbf{q}_p(t) = f_z^q \mathbf{Q}(t) \tag{5.103}$$

with

$$\mathbf{Q}(t) = \begin{pmatrix} \sum_{j=1}^{N} a_p g_j(t) \left(K_t \cos \varphi_j(t) + K_r \sin \varphi_j(t) \right) \left(f_z \sin \varphi_j(t) \right)^q \\ \sum_{j=1}^{N} a_p g_j(t) \left(-K_t \sin \varphi_j(t) + K_r \cos \varphi_j(t) \right) \left(f_z \sin \varphi_j(t) \right)^q \end{pmatrix} . \tag{5.104}$$

Equation (5.97) can be written in the first-order form

$$\dot{\mathbf{x}}(t) = \mathbf{A}(t)\mathbf{x}(t) + \mathbf{B}(t)\mathbf{u}(t - \tau) , \tag{5.105}$$

$$\mathbf{u}(t) = \mathbf{D}\mathbf{x}(t) , \tag{5.106}$$

where

$$\mathbf{x}(t) = \begin{pmatrix} \boldsymbol{\epsilon}(t) \\ \dot{\boldsymbol{\epsilon}}(t) \end{pmatrix} , \qquad \mathbf{D} = \begin{pmatrix} \mathbf{I} & \mathbf{0} \end{pmatrix} , \tag{5.107}$$

$$\mathbf{A}(t) = \begin{pmatrix} \mathbf{0} & \mathbf{I} \\ -\left(\mathbf{M}^{-1} \left(\mathbf{K} + \mathbf{G}(t) \right) \right) & -\mathbf{M}^{-1}\mathbf{C} \end{pmatrix} , \qquad \mathbf{B}(t) = \begin{pmatrix} \mathbf{0} \\ \mathbf{M}^{-1}\mathbf{G}(t) \end{pmatrix} , \tag{5.108}$$

with $\mathbf{I}$ denoting the 2×2 identity matrix. Stability charts for (5.105)–(5.106) can be determined by the semi-discretization method, as shown in Section 3.3.

Figure 5.21 shows the stability charts for up-milling operations with different radial immersion ratios a_e/D. The parameters are as follows. The number of cutting teeth is $N = 2$; the cutting-force coefficients are $K_t = 107 \times 10^6 \, \text{N/m}^{1+q}$ and $K_r = 40 \times 10^6 \, \text{N/m}^{1+q}$; the cutting-force exponent is $q = 0.75$. The feed per tooth is $f_z = 0.1 \, \text{mm}$, for which the linearized cutting-force coefficients are $K_t q f_z^{q-1} = 800 \times 10^6 \, \text{N/m}^2$ and $K_r q f_z^{q-1} = 300 \times 10^6 \, \text{N/m}^2$. The modal matrices are

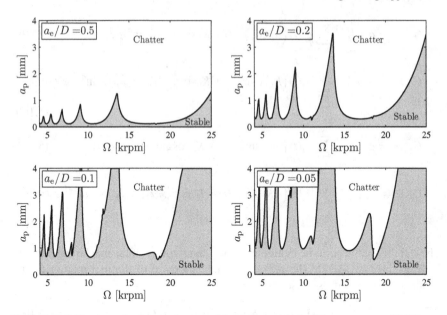

Fig. 5.21 Stability charts for the two-degrees-of-freedom model of milling processes with different radial immersions.

$$\mathbf{M} = \begin{pmatrix} 0.04 & 0 \\ 0 & 0.04 \end{pmatrix} \text{[kg]} \ , \quad \mathbf{C} = \begin{pmatrix} 8 & 0 \\ 0 & 8 \end{pmatrix} \left[\frac{\text{Ns}}{\text{m}} \right] , \quad \mathbf{K} = \begin{pmatrix} 1300 & 0 \\ 0 & 1300 \end{pmatrix} \left[\frac{\text{kN}}{\text{m}} \right] . \quad (5.109)$$

The corresponding natural frequency and damping ratio for both x and y directions are $f_n = 907$ Hz and $\zeta = 0.018$, respectively.

5.3 Periodic Control Systems with Feedback Delay

The main problem in the stabilization of control systems in the presence of feedback delay is that infinitely many characteristic exponents (poles) should be controlled, while the number of control parameters is finite. One technique to deal with the problem is to assign the place of the rightmost poles only while monitoring the other uncontrolled poles with large (negative) real part (see, e.g., Michiels et al. [188, 189]). This technique requires the numerical calculation of some relevant poles for different control parameters. In this case, the infinitely many poles are controlled by a finite number of control parameters.

An alternative approach is to increase the number of control parameters with the application of distributed delays in the controller, where the kernel function of the distributed delay serves as a kind of infinite-dimensional vector of control gains. Another way is the application of time-periodic controllers, where the time-dependency of the gains can be assumed to be a set of infinitely many control parameters.

A special case for distributed delay applications is that in which the feedback is based on a prediction of the state. This method is called finite spectrum assignment, since the resulting closed-loop system has only a finite number of poles that can be assigned arbitrarily, provided that there is no uncertainty in the system and in the control parameters (for details, see Manitius and Olbrot [178], Wang et al. [294]). If the open-loop system is unstable, then prediction by means of the solution of the differential equation cannot stabilize the system, since it involves an unstable pole–zero cancellation even for high-accuracy solutions (see Engelborghs et al. [75]). The conditions for stabilizing via distributed delays that approximates the solution of the system were analyzed by Mondié et al. [196], and a safe implementation of finite spectrum assignment for unstable systems was provided in Mondié and Michiels [195].

For delay-free systems, stabilization by means of periodic control gains is a field of intensive research (see, for instance, [36, 169, 199, 3, 31]). However, the combined effect of feedback delay and time-periodic control gains results in time-periodic DDEs, for which the stability analysis requires the use of the infinite-dimensional Floquet theory.

Actually, sampling can also be considered a special case of periodic controllers, since it corresponds to a periodic variation of the feedback delay in time, as was explained in Section 3.1 (see Figure 3.1). Other, generalized, sampled-data hold functions can also effectively be used to improve control performance (see, e.g,. Kabamba [144]). A special case of generalized hold discrete-time control is the intermittent predictive control, where the sequence of open-loop trajectories is punctuated by intermittent feedback. This concept was introduced by Ronco et al. [224] and further developed by Gawthrop and Wang [87, 88].

In this section, a special case of periodic controllers, the act-and-wait controller, is considered, where the feedback term is switched on and off periodically in time. The technique was introduced by Insperger and Stepan [118, 261, 127] for both continuous-time and discrete-time systems. The merit of the technique is that if the switch-off (waiting) period is longer than the feedback delay, then the system can be transformed to a discrete map of finite dimension presenting a finite spectrum assignment problem. This feature may be useful during the controller design. Several examples show that a stable control process can be attained by the application of the act-and-wait concept for problems where the traditional (constant gain) controllers cannot stabilize the system. Furthermore, the act-and-wait controller typically can be tuned to have dead-beat behavior (see, e.g., [118, 261, 155]). Due to its intermittent nature, the act-and-wait concept is relevant in biomechanical applications such as controlling biological networks [216] and human balancing [191, 13]. As was shown by Gawthrop [86], the act-and-wait controller is related to the intermittent controller in the sense that both techniques have a generalized hold interpretation.

5.3.1 The Act-and-Wait Control Concept

Consider the linear multiple-input multiple-output system

$$\dot{\mathbf{x}}(t) = \mathbf{A}\mathbf{x}(t) + \mathbf{B}\mathbf{u}(t) , \tag{5.110}$$

$$\mathbf{y}(t) = \mathbf{C}\mathbf{x}(t) , \tag{5.111}$$

with state $\mathbf{x}(t) \in \mathbb{R}^n$, input $\mathbf{u}(t) \in \mathbb{R}^m$, and output $\mathbf{y}(t) \in \mathbb{R}^l$. Consider the delayed feedback controller

$$\mathbf{u}(t) = \mathbf{D}\mathbf{y}(t - \tau) , \tag{5.112}$$

with τ being the feedback delay. It is assumed that the delay is a fixed parameter of the control system and cannot be eliminated or tuned during the control design. There are several sources of such time delays, e.g., acquisition of response and excitation data, information transmission, on-line data processing, computation and application of control forces. System (5.110)–(5.111) with (5.112) forms an autonomous DDE of the form

$$\dot{\mathbf{x}}(t) = \mathbf{A}\mathbf{x}(t) + \mathbf{B}\mathbf{D}\mathbf{C}\mathbf{x}(t - \tau) . \tag{5.113}$$

Stabilization of (5.113) brings the following pole placement problem: for given matrices $\mathbf{A}$, $\mathbf{B}$, and $\mathbf{C}$ and for given feedback delay τ, matrix $\mathbf{D}$ should be determined such that all the characteristic exponents (the poles) lie in the left half of the complex plane. The difficulty in this problem is that infinitely many characteristic exponents should be controlled by a finite number of control parameters, i.e., by the elements of matrix $\mathbf{D}$.

Consider now the periodic controller

$$\mathbf{u}(t) = \mathbf{G}(t)\mathbf{y}(t - \tau) , \tag{5.114}$$

where $\mathbf{G}(t)$ is a T-periodic function. System (5.110)–(5.111) with (5.114) implies the time-periodic DDE

$$\dot{\mathbf{x}}(t) = \mathbf{A}\mathbf{x}(t) + \mathbf{B}\mathbf{G}(t)\mathbf{C}\mathbf{x}(t - \tau) . \tag{5.115}$$

As was mentioned in Section 1.4, the general solution of (5.115) for the initial function $\mathbf{x}_0$ can be formulated as

$$\mathbf{x}_t = \mathcal{U}(t)\mathbf{x}_0 , \tag{5.116}$$

where $\mathcal{U}(t)$ is the solution operator (infinitesimal generator) of the system, and the function $\mathbf{x}_t$ is defined as

$$\mathbf{x}_t(\vartheta) = \mathbf{x}(t + \vartheta) , \quad \vartheta \in [-\tau, 0] . \tag{5.117}$$

The stability properties of the system are determined by the monodromy operator $\mathcal{U}(T)$. The nonzero elements of the spectrum of $\mathcal{U}(T)$ are called characteristic multipliers (or poles). The system is asymptotically stable if all the characteristic

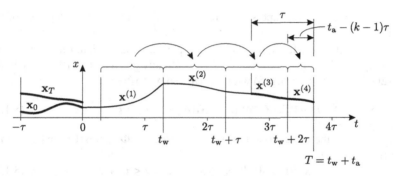

Fig. 5.22 Sketch of the piecewise solution of (5.115) with (5.118) according to the method of steps for $t_w \geq \tau$ and $2\tau < t_a \leq 3\tau$ ($k = 3$).

multipliers have modulus less than one. Like the time-independent system (5.113), stabilization of system (5.115) requires control over infinitely many poles, but now the control gains are functions of time over the period $[0, T]$, which corresponds to a kind of infinite-dimensional vector of control gains. The question is how to place all the infinitely many poles (i.e., infinitely many characteristic multipliers in this case) of the system inside the unit circle of the complex plane by tuning the control gain function $\mathbf{G}(t)$ over $[0, T]$. One possible approach to the problem is to decrease the dimension of the system using the act-and-wait control technique.

The act-and-wait controller is a special case of the time-varying controller (5.114) with the T-periodic matrix

$$\mathbf{G}(t) = \begin{cases} \mathbf{0} & \text{if } 0 \leq (t \bmod T) < t_w \,, \\ \mathbf{\Gamma}(t \bmod T) & \text{if } t_w \leq (t \bmod T) < t_w + t_a = T \,, \end{cases} \tag{5.118}$$

where $\mathbf{\Gamma} : [t_w, T] \to \mathbb{R}^{m \times l}$ is an integrable matrix function. Here, t_a and t_w are the lengths of the acting and the waiting periods, respectively, and $t_a + t_w = T$ is the length of one act-and-wait period.

In what follows, it will be shown that if the waiting period is larger than the feedback delay (i.e., if $t_w \geq \tau$), then the system can be described by an $n \times n$ monodromy matrix. Consequently, only n poles determine the stability instead of infinitely many.

Consider the general case $(k - 1)\tau < t_a \leq k\tau$, where k is an arbitrary positive integer. According to the method of steps for DDEs, the solution can be constructed piecewise over the consecutive intervals $[0, t_w], [t_w, t_w + \tau], \ldots, [t_w + (k - 1)\tau, T]$ step by step (see Figure 5.22 for $k = 3$).

Since the delayed term is switched off during the waiting period, the first section of the solution can be given as

$$\mathbf{x}^{(1)}(t) = \mathbf{\Phi}^{(1)}(t)\mathbf{x}(0) \,, \qquad 0 \leq t \leq t_w \,, \tag{5.119}$$

with $\mathbf{\Phi}^{(1)}(t) = e^{\mathbf{A}t}$. Here, superscript (1) refers to the number of the segment of the solution.

Now we utilize the facts that the waiting period is larger than (or equal to) the time delay and the solution over $0 \leq t \leq t_w$ is already given by (5.119). Thus, in the interval $t_w < t \leq t_w + \tau$, (5.115) with (5.118) can be written as

$$\dot{\mathbf{x}}(t) = \mathbf{A}\mathbf{x}(t) + \mathbf{B}\mathbf{\Gamma}(t)\mathbf{C}\mathbf{\Phi}^{(1)}(t - \tau)\mathbf{x}(0) , \qquad t_w < t \leq t_w + \tau . \qquad (5.120)$$

The solution for the initial condition $\mathbf{x}(t_w) = \mathbf{x}^{(1)}(t_w) = \mathbf{\Phi}^{(1)}(t_w)\mathbf{x}(0)$ can be given in the form

$$\mathbf{x}^{(2)}(t) = \mathbf{\Phi}^{(2)}(t)\mathbf{x}(0) , \qquad t_w < t \leq t_w + \tau , \qquad (5.121)$$

with

$$\mathbf{\Phi}^{(2)}(t) = e^{\mathbf{A}t} + \int_{t_w}^{t} e^{\mathbf{A}(t-s)}\mathbf{B}\mathbf{\Gamma}(s)\mathbf{C}\mathbf{\Phi}^{(1)}(s - \tau)\, ds \qquad (5.122)$$

(see the variation of constants formula (A.15) in Appendix A.1). Provided that the solution in the hth interval is given as

$$\mathbf{x}^{(h)}(t) = \mathbf{\Phi}^{(h)}(t)\mathbf{x}(0) , \qquad t_w + (h - 2)\tau < t \leq t_w + (h - 1)\tau , \qquad (5.123)$$

the solution in the next interval can be given by the recursive form

$$\mathbf{x}^{(h+1)}(t) = \mathbf{\Phi}^{(h+1)}(t)\mathbf{x}(0) , \qquad t_w + (h - 1)\tau < t \leq t_w + h\tau , \qquad (5.124)$$

with

$$\mathbf{\Phi}^{(h+1)}(t) = e^{\mathbf{A}t} + \int_{t_w-(h-1)\tau}^{t} e^{\mathbf{A}(t-s)}\mathbf{B}\mathbf{\Gamma}(s)\mathbf{C}\mathbf{\Phi}^{(h)}(s - \tau)\, ds . \qquad (5.125)$$

Finally, the solution at $t = T$ is given as

$$\mathbf{x}(T) = \mathbf{x}^{(k+1)}(T) = \mathbf{\Phi}^{(k+1)}(T)\mathbf{x}(0) . \qquad (5.126)$$

As can be seen, the state $\mathbf{x}(T)$ depends only on the initial state $\mathbf{x}(0)$, and it does not depend on the initial function $\mathbf{x}_0$. Matrix $\mathbf{\Phi}^{(k+1)}(T)$ therefore serves as an $n \times n$ monodromy matrix for the system.

As a consequence, the infinite-dimensional monodromy mapping

$$\mathbf{x}_T = \mathcal{U}(T)\mathbf{x}_0 \qquad (5.127)$$

can be written in the form

$$\begin{pmatrix} \mathbf{x}(T) \\ \tilde{\mathbf{x}}_T \end{pmatrix} = \begin{pmatrix} \mathbf{\Phi}^{(k+1)}(T) & \mathbf{O} \\ \tilde{\mathbf{f}}_{k+1} & O \end{pmatrix} \begin{pmatrix} \mathbf{x}(0) \\ \tilde{\mathbf{x}}_0 \end{pmatrix} , \qquad (5.128)$$

where the function $\tilde{\mathbf{x}}_t$ is defined by the shift

$$\tilde{\mathbf{x}}_t(\vartheta) = \mathbf{x}(t + \vartheta) , \qquad \vartheta \in [-\tau, 0) . \qquad (5.129)$$

Note that $\vartheta = 0$ is excluded here as opposed to x_t in (5.117). In (5.128), $\mathbf{O}$ denotes the zero functional, O denotes the zero operator, and $\tilde{\mathbf{f}}_{k+1}$ is the function

$$\tilde{\mathbf{f}}_{k+1}(\vartheta) = \begin{cases} \mathbf{\Phi}^{(k)}(\vartheta + T) & \text{if } -\tau \leq \vartheta < (k-1)\tau - t_a , \\ \mathbf{\Phi}^{(k+1)}(\vartheta + T) & \text{if } (k-1)\tau - t_a \leq \vartheta < 0 . \end{cases} \tag{5.130}$$

Equation (5.128) shows that the function $\mathbf{x}_T$ can be determined using only the initial state $\mathbf{x}(0)$ and does not depend on the initial function $\tilde{\mathbf{x}}_0$. The stability properties are determined by the $n \times n$ matrix $\mathbf{\Phi}^{(k+1)}(T)$. The system is asymptotically stable if all the eigenvalues of $\mathbf{\Phi}^{(k+1)}(T)$ have modulus less than 1. Therefore, in this case, the stability analysis requires the calculation of the eigenvalues of the $n \times n$ matrix $\mathbf{\Phi}^{(k+1)}(T)$ only.

For instance, if $k = 1$, i.e., $0 < t_a \leq \tau$, then the monodromy matrix is given as

$$\mathbf{\Phi}^{(2)}(T) = e^{\mathbf{A}T} + \int_{t_w}^T e^{\mathbf{A}(T-s)}\mathbf{B\Gamma}(s)\mathbf{C}e^{\mathbf{A}(s-\tau)}\,ds . \tag{5.131}$$

If $k = 2$, i.e., $\tau < t_a \leq 2\tau$, then

$$\mathbf{\Phi}^{(3)}(T) = e^{\mathbf{A}T} + \int_{t_w}^T e^{\mathbf{A}(T-s)}\mathbf{B\Gamma}(s)\mathbf{C}e^{\mathbf{A}(s-\tau)}\,ds$$

$$+ \int_{t_w+\tau}^T e^{\mathbf{A}(T-s_1)}\mathbf{B\Gamma}(s_1)\mathbf{C}\int_{t_w}^{s_1-\tau} e^{\mathbf{A}(s_1-s_2-\tau)}\mathbf{B\Gamma}(s_2)\mathbf{C}e^{\mathbf{A}(s_2-\tau)}\,ds_2\,ds_1 . \tag{5.132}$$

Note that if the waiting period is less then the feedback delay, i.e., $t_w < \tau$, then the above derivation is not valid, and the monodromy operator cannot be represented in the $n \times n$ matrix form as in (5.131) or (5.132). Still, if the waiting and the acting periods t_w and t_a satisfy certain conditions, then a finite-dimensional monodromy matrix can be constructed for the case $t_w < \tau$, too (for details, see [128]).

5.3.2 Stick-Balancing with Reflex Delay

Balancing an inverted pendulum in the presence of feedback delay is a frequently cited example in dynamics and control theory [255, 243, 166], and it is also a relevant issue to human motion control [24, 190, 192, 13, 173]. It is known that conventional proportional-derivative (PD) controllers cannot stabilize the upward position if the time delay is larger than a critical value. As was shown by Stepan [260], this critical delay for a continuous PD feedback can be given in the simple form $\tau_{\text{crit}} = T_p/(\pi\sqrt{2})$, where T_p is the period of the small oscillations of the pendulum hanging at its downward position. The same phenomenon is often communicated such that for a given feedback delay, there is a critical minimum length of the pen-

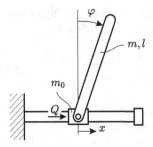

Fig. 5.23 Mechanical model of stick-balancing.

dulum: if the pendulum is shorter than this critical length, then the upward position is unstable for any PD controller [47].

Here, it will be shown that the application of the act-and-wait control concept helps in the stabilization of the upper equilibrium of the pendulum. The mechanical model of the system can be seen in Figure 5.23. The pendulum of length l and mass m is attached to the horizontal slide. The mass m_0 of the slide is assumed to be negligible relative to the mass of the pendulum. The general coordinates are the angular position φ of the stick and the position x of the pivot point. A control force Q is applied on the slide in order to balance the stick at $\varphi = 0$. The equation of motion for this system takes the form

$$\begin{pmatrix} \frac{1}{3}ml^2 & \frac{1}{2}ml\cos\varphi \\ \frac{1}{2}ml\cos\varphi & m \end{pmatrix}\begin{pmatrix} \ddot{\varphi} \\ \ddot{x} \end{pmatrix} + \begin{pmatrix} -\frac{1}{2}mgl\sin\varphi \\ -\frac{1}{2}ml\dot{\varphi}^2\sin\varphi \end{pmatrix} = \begin{pmatrix} 0 \\ Q(\varphi,\dot{\varphi}) \end{pmatrix}, \tag{5.133}$$

where g stands for the gravitational acceleration. The displacement x is a cyclic coordinate that can be eliminated from the equation. The essential motion φ is then governed by

$$\left(\frac{1}{3}ml^2 - \frac{1}{4}ml^2\cos^2\varphi\right)\ddot{\varphi} + \frac{1}{8}ml^2\dot{\varphi}^2\sin(2\varphi) - \frac{1}{2}mgl\sin\varphi = -\frac{1}{2}lQ(\varphi,\dot{\varphi})\cos\varphi. \tag{5.134}$$

The control force Q is assumed to be a locally linear function of the angular position φ and the angular velocity $\dot{\varphi}$ in the form

$$Q(\varphi,\dot{\varphi}) = K_p\varphi + K_d\dot{\varphi} + \text{h.o.t.}, \tag{5.135}$$

where K_p is the proportional gain, K_d is the derivative gain, and h.o.t. stands for the higher-order terms not modeled here.

Linearization around the upright position $\varphi = 0$ and modeling the delay τ in the feedback loop gives

$$\frac{1}{12}ml^2\ddot{\varphi}(t) - \frac{1}{2}mgl\varphi(t) = -\frac{1}{2}l\left(K_p\varphi(t-\tau) + K_d\dot{\varphi}(t-\tau)\right). \tag{5.136}$$

Introducing new parameters, the system can be transformed into the form

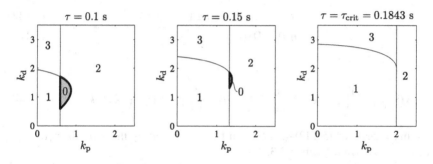

Fig. 5.24 Number of unstable characteristic exponents for (5.137) with length $l = 1$ m and with different feedback delays τ.

$$\ddot{\varphi}(t) + a_0\varphi(t) = -k_p\varphi(t - \tau) - k_d\dot{\varphi}(t - \tau) , \qquad (5.137)$$

where

$$a_0 = -\frac{6g}{l} < 0 , \qquad k_p = \frac{6K_p}{ml} , \qquad k_d = \frac{6K_d}{ml} . \qquad (5.138)$$

This equation is equivalent to (2.49), for which the stability properties were analyzed in Section 2.3 and the stability charts for different parameters are given in Figures 2.7 and 2.8. The same diagrams are shown in Figure 5.24 for pendulum length $l = 1$ m and for different feedback delays τ. In fact, these diagrams are the reparameterizations of those shown in Figure 2.8. Here, the stable domain shrinks with increasing feedback delay τ, and it disappears if $\tau > \tau_{crit}$. According to (2.62), the value of the critical delay can be given as

$$\tau_{crit} = \sqrt{\frac{-2}{a_0}} = \sqrt{\frac{l}{3g}} = \frac{T_p}{\pi\sqrt{2}} , \qquad (5.139)$$

where $T_p = \pi\sqrt{2l/(3g)}$ is the period of the small oscillations of the pendulum about its downward equilibrium. For a pendulum of length $l = 1$ m, PD controllers with feedback delays larger than $\tau_{crit} = 0.1843$ s cannot stabilize the upward position.

Consider now the same system subjected to the act-and-wait controller in the form

$$\ddot{\varphi}(t) + a_0\varphi(t) = -g(t)\left(k_p\varphi(t - \tau) + k_d\dot{\varphi}(t - \tau)\right) , \qquad (5.140)$$

where

$$g(t) = \begin{cases} 0 & \text{if } 0 \le t \bmod T < t_w , \\ 1 & \text{if } t_w \le t \bmod T < t_w + t_a = T , \end{cases} \qquad (5.141)$$

is the act-and-wait switching function. Here, the feedback is zero for the waiting period, and constant proportional and derivative gains are applied in the acting period. The system can be written in the form

$$\dot{x}(t) = \mathbf{A}x(t) + g(t)\mathbf{B}u(t - \tau) , \tag{5.142}$$

$$\mathbf{u}(t) = \mathbf{D}x(t) , \tag{5.143}$$

where

$$x(t) = \begin{pmatrix} \varphi(t) \\ \dot{\varphi}(t) \end{pmatrix} , \quad \mathbf{A} = \begin{pmatrix} 0 & 1 \\ -a_0 & 0 \end{pmatrix} , \quad \mathbf{B} = \begin{pmatrix} 0 \\ 1 \end{pmatrix} , \quad \mathbf{D} = \begin{pmatrix} -k_p & -k_d \end{pmatrix} . \tag{5.144}$$

Stability charts for (5.142)–(5.143) can be determined by the semi-discretization method, as shown in Section 3.3.

For the special case $t_w \geq \tau$, the monodromy matrix can also be given in closed form, as was shown in Section 5.3.1. For this purpose, the system should first be written in the form of (5.115), where $\mathbf{C}$ is the 2×2 identity matrix and $\mathbf{G}(t) = g(t)\mathbf{D}$. Then, the recurrence (5.125) with (5.122) can be used to obtain the closed-form representation of the monodromy matrix (see, e.g., (5.131) or (5.132)).

A measure for the decay of the oscillations around the upper equilibrium is characterized by the magnitude of the critical (largest in modulus) characteristic multiplier μ_1, i.e., $\|x_{t+T}\| \leq |\mu_1| \|x_t\|$. In order to compare cases with different act-and-wait periods $T = t_w + t_a$, introduce the decay ratio such that $\rho = |\mu_1|^{1/T}$. This decay ratio characterizes the decay over a unit time step, i.e., $\|x_{t+1}\| \leq \rho \|x_t\|$.

Figure 5.25 shows a series of stability charts in the plane (k_p, k_d) for a pendulum of length $l = 1\,m$ with feedback delay $\tau = 0.1\,s$ for different acting and waiting periods. These diagrams can be considered projections of the 4 dimensional stability chart in the parameter space (k_p, k_d, t_a, t_w). The contour lines where the decay ratio ρ is equal to $1, 1.5, 2, \ldots$ are also presented. The stability boundaries, where $\rho = 1$, are indicated by thick lines. It can be seen that there is a qualitative change in the structure of the stable domains if the waiting period t_w becomes larger than the feedback delay τ. The reason is that in this case, the dynamics changes radically: the dimension of the system is reduced to $n = 2$, and the system is described by the 2×2 monodromy matrix. Consequently, the stability diagrams for the case $t_w \geq \tau = 0.1\,s$ can also be obtained by analyzing the 2×2 matrix $\mathbf{\Phi}^{(2)}(T)$ or $\mathbf{\Phi}^{(3)}(T)$ in (5.131) or (5.132) depending on whether $0 < t_a \leq \tau$ or $\tau < t_a \leq 2\tau$.

In Figure 5.25, the feedback delay ($\tau = 0.1\,s$) is smaller than the critical delay $\tau_{crit} = 0.1843\,s$; consequently, the system can be stabilized by the traditional (constant gain) controller as well. The diagrams for the case $t_w = 0$ correspond in fact to the stability chart in the left panel of Figure 5.24.

The same diagrams are presented in Figure 5.26 for a pendulum of length $l = 1\,m$ with feedback delay $\tau = 0.2\,s$. In this case, the feedback delay is larger than the critical value; consequently, the system cannot be stabilized by constant control gains. Periodic switching of the control gains according to the act-and-wait concept may, however, result in a stabilizing control. As can be seen, large triangular stable domains appear for $t_w \geq \tau = 0.2$, but small stable domains can also be observed in some plots with $t_w = 0.15\,s$.

The act-and-wait concept provides an alternative for control systems with feedback delays. The traditional approach is the continuous use of constant control gains,

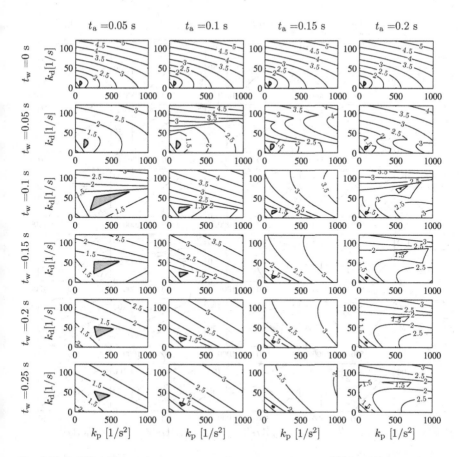

Fig. 5.25 Stability charts and contour curves of the decay ratio $\rho = |\mu_1|^{1/T}$ for different acting and waiting periods for a pendulum of length $l = 1$ m with feedback delay $\tau = 0.1$ s. Stable domains are indicated by gray shading.

when a cautious, slow feedback is applied with small gains, resulting in slow convergence (if such a controller can stabilize the system at all). The act-and-wait control concept is a special case of periodic controllers, where time-varying control gains are used in the acting phase and zero gains are used in the waiting intervals. Several (actually, infinitely many) periodic functions could be chosen as time-periodic controllers. The main idea behind choosing the one that involves waiting intervals just longer than the feedback delay is that this kills the memory effect by waiting for the system's response induced by the previous action. Although it might seem unnatural not to actuate during the wait interval at all, the act-and-wait concept is still a natural control logic for time-delayed systems. This is how, for example, one would adjust the shower temperature considering the delay between the controller (tap) and the sensed output (water temperature at skin).

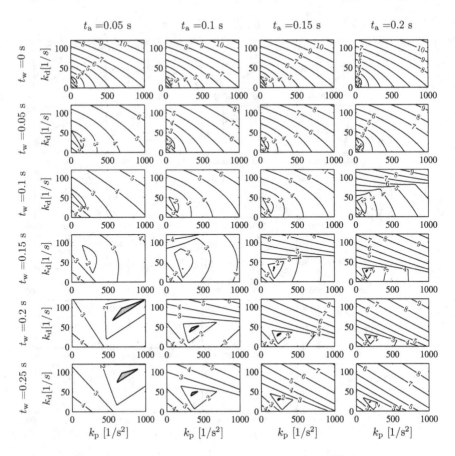

Fig. 5.26 Stability charts and contour curves of the decay ratio $\rho = |\mu_1|^{1/T}$ for different acting and waiting periods for a pendulum of length $l = 1$ m with feedback delay $\tau = 0.2$ s. Stable domains are indicated by gray shading.

5.3.3 Force Control

Force control is an essential mechanical controlling problem in robotics, since most robotic applications involve interactions with other objects. The first publications on the basics of force-control approaches appeared in the early 1980s, starting with the pioneering work of Whitney [295], Mason [182], and Raibert and Craig [222]. Since then, several comprehensive textbooks have been published summarizing different methods of force-control processes in the field of robotics [57, 12, 50]. The aim of force control is to provide a desired force between the actuator and the environment (or workpiece). In order to achieve high accuracy in maintaining the prescribed contact force against Coulomb friction, high proportional control gains are to be used [57, 12]. However, in practical realizations of force-control processes

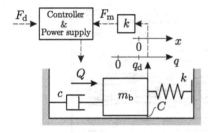

Fig. 5.27 Single-degree-of-freedom mechanical model of force-control process.

with high proportional gains, the robot often loses stability, and starts to oscillate at a relatively low frequency. These oscillations are mainly caused by digital effects [295, 264, 258, 157, 156] and by time delays in the feedback loop [255, 89]. In spite of efforts to minimize time delays, they cannot be eliminated totally, even with today's advanced technology due to physical limits. Teleoperation is a typical example in which communication delay plays a crucial role [150, 9, 200, 220], but similar delays may arise in haptic interfaces as well [52].

The single-degree-of-freedom mechanical model of the force-control process is shown in Figure 5.27. The modal mass m_b and the equivalent stiffness k represent the inertia and the stiffness of the robot and the environment, while equivalent damping c models the viscous damping due to the servo motor characteristics and the environment. The force Q represents the controller's action, and C is the magnitude of the effective Coulomb friction. Considering a proportional force controller, the control force can be given as

$$Q(t) = F_d - k_p \left(F_m(t) - F_d \right) , \qquad (5.145)$$

where k_p is the proportional gain, F_d is the desired force, and F_m is the measured force. The equation of motion reads

$$m_b \ddot{q}(t) + c\dot{q}(t) + kq(t) = F_d - k_p \left(F_m(t) - F_d \right) - C\mathrm{sgn}(\dot{q}(t)) . \qquad (5.146)$$

Assuming a steady-state condition by setting all the time derivatives to zero, considering a constant Coulomb friction force, and using that $F_m = kq(t)$, the maximum force error can be given as

$$F_e^{max} = \frac{C}{1 + k_p} . \qquad (5.147)$$

Thus, the higher the gain k_p, the less the force error. Theoretically, there is no upper limit for the gain k_p, since the constant solution $q(t) \equiv q_d$ of (5.146) is always asymptotically stable when $C = 0$. Experiments show, however, that the real system with feedback delay is not stable for large gain k_p [264].

In practical realizations, the control force can be written in the form

$$Q(t) = F_d - k_p(F_m(t - \tau) - F_d) = kq_d - k_p \left(kq(t - \tau) - kq_d \right) , \qquad (5.148)$$

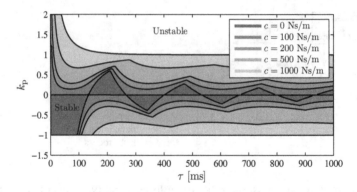

Fig. 5.28 Stability charts for (5.150) with different damping parameters.

where τ is the time delay in the feedback loop. Thus, the equation of motion reads

$$m_b\ddot{q}(t) + c\dot{q}(t) + kq(t) = kq_d - k_p\left(kq(t-\tau) - kq_d\right) - C\text{sgn}(\dot{q}(t)) . \qquad (5.149)$$

Stability properties of the system can be determined by analyzing the variational system of (5.149) around the desired position q_d. For this calculation, the dry friction is neglected in the model. Considering that $q(t) = q_d + x(t)$, the variational system reads

$$\ddot{x}(t) + 2\zeta\omega_n\dot{x}(t) + \omega_n^2 x(t) = -\omega_n^2 k_p x(t-\tau) , \qquad (5.150)$$

where $\omega_n = \sqrt{k/m_b}$ is the natural angular frequency of the uncontrolled undamped system and $\zeta = c/(2m_b\,\omega_n)$ is the damping ratio. Equation (5.150) is actually a reparameterization of the delayed oscillator (2.28), and the stability analysis can be performed by the D-subdivision method in a way similar to that shown in Section 2.2. The stability boundaries in the plane (τ, k_p) can be given as

$$\text{if } \omega = 0 : \quad k_p = -1 , \qquad (5.151)$$

$$\text{if } \omega \neq 0 : \quad \tau = \frac{1}{\omega}\left(j\pi - \arctan\left(\frac{2\zeta\omega_n\omega}{\omega_n^2 - \omega^2}\right)\right) , \quad j \in \mathbb{Z} , \qquad (5.152)$$

$$k_p = \frac{(-1)^j\text{sgn}(\omega - \omega_n)}{\omega_n^2}\sqrt{(\omega_n^2 - \omega^2)^2 + 4\zeta^2\omega_n^2\omega^2} . \qquad (5.153)$$

Stability charts for different damping parameters c are shown in Figure 5.28. The mass and the stiffness parameters are $m_b = 29.57\,\text{kg}$ and $k = 16414\,\text{N/m}$. Stable domains with different gray shades are associated with different damping parameters c. If the feedback delay is zero, then the system is asymptotically stable for any $k_p > -1$. Note that the system without control (i.e., if $k_p = 0$) is stable itself, and the goal of the control is to ensure an accurate contact force.

Consider now the same system with an act-and-wait controller. In this case, the control force can be given as

$$Q_{a\&w}(t) = F_d - g(t)k_p(F_m(t - \tau) - F_d) , \qquad (5.154)$$

where $g(t)$ is the T-periodic act-and-wait switching function defined in (5.141). Thus,

$$Q_{a\&w} = \begin{cases} F_d & \text{if } 0 \le t \bmod T < t_w , \\ F_d - k_p(F_m(t - \tau) - F_d) & \text{if } t_w \le t \bmod T < t_w + t_a = T . \end{cases} \qquad (5.155)$$

This means that the control force is equal to the desired force for period of length t_w, and the feedback is switched on only for periods of length t_a. The corresponding variational system reads

$$\ddot{x}(t) + 2\zeta\omega_n\dot{x}(t) + \omega_n^2 x(t) = -g(t)\omega_n^2 k_p x(t - \tau) . \qquad (5.156)$$

Transformation into first-order form gives

$$\dot{\mathbf{x}}(t) = \mathbf{A}\mathbf{x}(t) + g(t)\mathbf{B}\mathbf{u}(t - \tau) , \qquad (5.157)$$
$$\mathbf{u}(t) = \mathbf{D}\mathbf{x}(t) , \qquad (5.158)$$

with

$$\mathbf{x}(t) = \begin{pmatrix} x(t) \\ \dot{x}(t) \end{pmatrix} , \quad \mathbf{A} = \begin{pmatrix} 0 & 1 \\ -\omega_n^2 & -2\zeta\omega_n \end{pmatrix} , \quad \mathbf{B} = \begin{pmatrix} 0 \\ -\omega_n^2 \end{pmatrix} , \quad \mathbf{D} = \begin{pmatrix} k_p & 0 \end{pmatrix} . \quad (5.159)$$

Stability charts for (5.157)–(5.158) can be determined by the semi-discretization method, as shown in Section 3.3. Alternatively, if $t_w \ge \tau$ and $0 < t_a \le \tau$, then the monodromy matrix of the system can also be determined in closed form according to (5.131).

Let μ_1 denote the critical (maximum in modulus) eigenvalue. The system is asymptotically stable if $|\mu_1| < 1$. The frequency of the resulting self-excited vibrations at the loss of stability is related to the phase angle

$$\omega_1 = \frac{1}{T}\text{Im} \left(\ln(\mu_1)\right) = \frac{1}{T} \arctan\left(\frac{\text{Im}\,\mu_1}{\text{Re}\,\mu_1}\right) \qquad (5.160)$$

with $-\pi < \omega_1 T \le \pi$. The vibration frequencies are the positive values of

$$f = \pm\frac{\omega_1}{2\pi} + \frac{j}{T} \quad [\text{Hz}], \quad j = 0, \pm 1, \pm 2, \ldots . \qquad (5.161)$$

Figure 5.29 presents the stability charts (middle panel), the vibration frequencies along the stability boundaries (top panel), and the maximum force error (bottom panel) for the continuous controller described by (5.148) and for the act-and-wait controller given by (5.155). The parameters are $m_b = 29.57\,\text{kg}$, $k = 16414\,\text{N/m}$, $c = 1447\,\text{Ns/m}$, and the Coulomb friction force is $C = 16.5\,\text{N}$. The length of the waiting period is equal to the feedback delay, i.e., $t_w = \tau$, while the ratio of the acting period length and the delay is set to a fixed number $t_a/\tau = 0.2$. It can be seen that

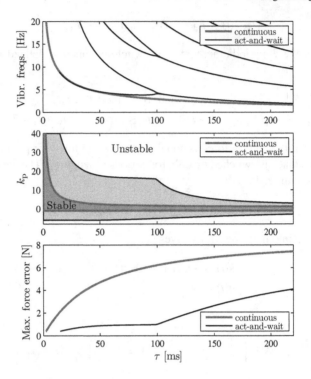

Fig. 5.29 Theoretically predicted stability charts, frequency diagrams and the maximum force error for the continuous and for the act-and-wait control concept.

the maximum achievable stable proportional gain k_p is larger for the act-and-wait controller than for the continuous one. The maximum force error is determined by the maximum stable gain k_p and (5.147). Consequently, the act-and-wait controller results in a smaller force error than the continuous controller. The frequency diagram shows that while the continuous control case is associated with a single vibration frequency, for the act-and-wait control case, a series of vibration frequencies arises according to (5.161).

Figures 5.30 and 5.31 show a comparison of the theoretical predictions with some experimental results for the continuous controller and for the act-and-wait controller, respectively. For the experimental results, a HIRATA (MB-H180-500) DC drive robot was used. The axis of the robot was connected to the base of the robot (environment) by a helical spring of stiffness $k = 16414\,\text{N/m}$. The contact force was measured by a Tedea-Huntleight Model 355 load cell mounted between the spring and the robot's flange. The driving system of the moving axis consisted of a HIRATA HRM-020-100-A DC servo motor connected directly to a ballscrew with a 20 mm pitch thread. The robot was controlled by a micro-controller based control unit providing the maximum sampling frequency 1 kHz for the overall force-control loop. This controller made it also possible to vary the time delay as integer multi-

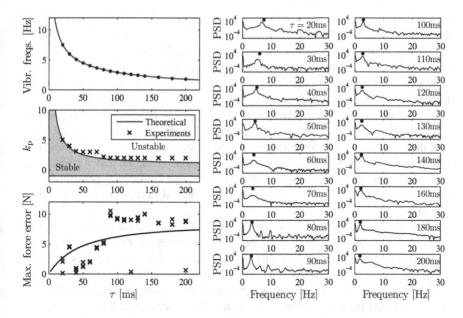

Fig. 5.30 Comparison of the theoretical stability chart, vibration frequencies, and force errors to the experimental results for the continuous controller.

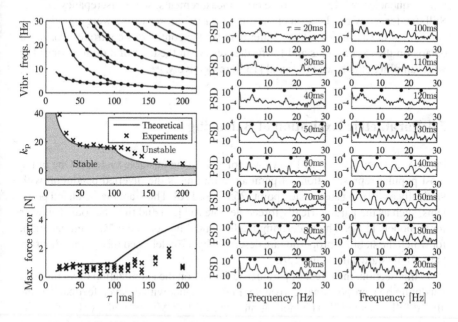

Fig. 5.31 Comparison of the theoretical stability chart, vibration frequencies, and force errors to the experimental results for the act-and-wait controller.

ples of 1 ms, and to set the control force by the pulse width modulation (PWM) of supply voltage of the DC motor. Time delay was varied between 20 and 200 ms, which are significantly larger than the sampling period 1 ms; therefore, the system can be considered a continuous-time system. The modal mass and the damping ratio were experimentally determined: $m_b = 29.57$ kg and $c = 1447$ Ns/m. The Coulomb friction was measured to be $C = 16.5$ N. The desired force was $F_d = 50$ N.

During the measurements, the time delay was fixed and the proportional gain was increased slowly until the process lost stability for perturbations larger than 50 N. The displacement of the force sensor was recorded during the loss of stability in order to analyze the frequency content of the motion. Then, the gain k_p was set to 90% of the critical value to obtain a stable process, the system was perturbed three times, and the resulting force errors were documented (three for each fixed time delay).

Figures 5.30 and 5.31 show the stability charts (left middle panel), the associated frequency diagrams (left top panel), the maximum force error (left bottom panel), and series of power spectrum density (PSD) diagrams for the test points with different feedback delays (right panels). The experimental stability boundaries and the measured maximum force errors are represented by crosses. The theoretically predicted vibration frequencies are also shown in the experimental PSD diagrams for reference. It can clearly be seen that the experimental results show good agreement with the theoretical predictions with respect to the stability boundaries and the vibration frequencies, while in the force error measurements, some discrepancy occurs. Still, the tendency can be seen clearly: the measured force errors were significantly smaller for the act-and-wait controller than for the continuous controller. For more details on the experiments and for further results in the case of a digital controller with sampling effect, see [134] and [135], respectively.

5.4 Parametric Excitation of Balancing

Parametric forcing is a well-known way of stabilizing unstable systems. One of the most popular representations of parametric forcing is the inverted pendulum with vertically oscillating suspension point [265, 146, 170]. There are many related problems, such as the parametrically forced flexible rod [51] and time-periodic follower force models [174]. The underlying mathematical model is the Mathieu equation, for which the stability properties are described by the celebrated Ince–Strutt diagram [116, 286] (shown in Figure 2.9).

In this section, the effect of parametric forcing on the stabilization of an inverted pendulum by a proportional-derivative (PD) controller with delayed feedback is analyzed following [120]. The motivation is the work of Milton et al. [191], who analyzed people's stick-balancing abilities, while standing on a vibrating platform. The mechanical model in question is shown in Figure 5.32. The stick is attached to the horizontal slide, which moves periodically up and down together with the base according to the geometric constraint $r_a \cos(\omega t)$. The stick to be balanced is assumed

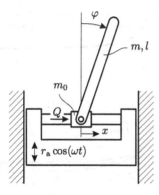

Fig. 5.32 Mechanical model of stick-balancing with parametric excitation.

to be homogeneous; its mass is m and its length is l. The mass m_0 of the slide is assumed to be negligible relative to the mass of the stick. The general coordinates are the angular position φ of the stick and the position x of the pivot point. A control force Q is applied to the slide in order to balance the stick. The equation of motion for this dynamical system takes the form

$$\begin{pmatrix} \frac{1}{3}ml^2 & \frac{1}{2}ml\cos\varphi \\ \frac{1}{2}ml\cos\varphi & m \end{pmatrix} \begin{pmatrix} \ddot{\varphi} \\ \ddot{x} \end{pmatrix} + \begin{pmatrix} \left(-\frac{1}{2}mgl + \frac{1}{2}mlr_a\omega^2\cos(\omega t)\right)\sin\varphi \\ -\frac{1}{2}ml\dot{\varphi}^2\sin\varphi \end{pmatrix} = \begin{pmatrix} 0 \\ Q(\varphi,\dot{\varphi}) \end{pmatrix} .$$
(5.162)

The displacement x is a cyclic coordinate that can be eliminated from the equation. The essential motion φ is then governed by

$$\left(\frac{1}{3}ml^2 - \frac{1}{4}ml^2\cos^2\varphi\right)\ddot{\varphi} + \frac{1}{8}ml^2\dot{\varphi}^2\sin(2\varphi)$$

$$+ \left(-\frac{1}{2}mgl + \frac{1}{2}mlr_a\omega^2\cos(\omega t)\right)\sin\varphi = -\frac{1}{2}lQ(\varphi,\dot{\varphi})\cos\varphi . \quad (5.163)$$

The control force Q is assumed to be a locally linear function of the angular position φ and the angular velocity $\dot{\varphi}$ in the form

$$Q(\varphi,\dot{\varphi}) = K_p\varphi + K_d\dot{\varphi} + \text{h.o.t.} , \tag{5.164}$$

where K_p is the proportional gain, K_d is the derivative gain, and h.o.t. stands for the higher-order terms not modeled here.

Linearization around the upright position $\varphi = 0$ and modeling the delay τ in the feedback loop gives

$$\frac{1}{12}ml^2\ddot{\varphi}(t) + \left(-\frac{1}{2}mgl + \frac{1}{2}mlr_a\omega^2\cos(\omega t)\right)\varphi(t) = -\frac{1}{2}l\left(K_p\varphi(t-\tau) + K_d\dot{\varphi}(t-\tau)\right) .$$
(5.165)

Introducing new parameters, the system can be transformed into the form

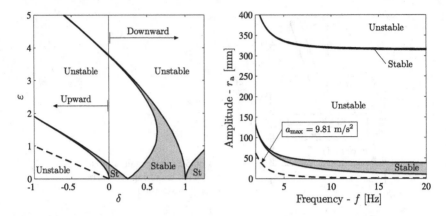

Fig. 5.33 Stability chart for (5.166) with $k_p = 0$ and $k_d = 0$ in the plane (δ, ε) (left) and in the plane of the forcing frequency f and amplitude r_a for a pendulum of length $l = 50\,\text{cm}$ (right). Dashed lines indicate the parameters for which the maximum acceleration of the pivot point is equal to g.

$$\ddot{\varphi}(t) + (\delta + \varepsilon \cos(\omega t))\,\varphi(t) = -k_p \varphi(t - \tau) - k_d \dot{\varphi}(t - \tau)\,, \qquad (5.166)$$

where

$$\delta = -\frac{6g}{l}\,, \qquad \varepsilon = \frac{6 r_a \omega^2}{l}\,, \qquad k_p = \frac{6 K_p}{ml}\,, \qquad k_d = \frac{6 K_d}{ml}\,. \qquad (5.167)$$

Equation (5.166) covers two special cases: (1) when the controller is switched off, i.e., $k_p = 0$ and $k_d = 0$; (2) when there is no parametric forcing, i.e., $r_a = 0$ or $\varepsilon = 0$.

Case (1):
If $k_p = 0$ and $k_d = 0$, then one obtains the classical Mathieu equation that describes the behavior of a pendulum under parametric forcing around the upward and downward positions. The stability diagram in the plane (δ, ε) (the Ince–Strutt diagram [116, 286]) can be seen in the left panel of Figure 5.33. The case $\delta > 0$ corresponds to the downward position of the pendulum, while the case $\delta < 0$ corresponds to the upward position (inverted pendulum). The diagram in the right panel of Figure 5.33 presents the stability domains transformed to the plane of the forcing frequency $f = \omega/(2\pi)$ and the forcing amplitude r_a for a pendulum of length $l = 50\,\text{cm}$. The case in which the maximum acceleration of the pivot point is equal to the gravitational acceleration g is indicated by a dashed line. The stable domains are located above this limit, i.e., the upward position of the pendulum can be stabilized by parametric forcing only if the maximum acceleration of the pivot point exceeds g.

Case (2):
If the amplitude of the parametric forcing is equal to zero, then one gets the governing equation for the PD control of an inverted pendulum with delayed feedback.

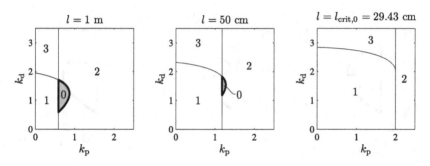

Fig. 5.34 Number of unstable characteristic exponents for (5.166) with $\varepsilon = 0$ for feedback delay $\tau = 0.1$ s and for different lengths l.

This case was analyzed in Section 2.3 with $\delta = a_0$ and $\varepsilon = 0$. The corresponding stability charts can be seen in Figures 2.7 and 2.8. The same diagrams are shown in Figure 5.34 for feedback delay $\tau = 0.1$ s and for different lengths l of the pendulum. Like Figure 5.24, this diagram is a reparameterization of those shown in Figures 2.8. Here, the stable domain shrinks with decreasing length l, and it disappears if $l < l_{\text{crit},0} = 3g\tau^2$ (see also (5.139) and the corresponding explanation in Section 5.3.2). For a feedback delay $\tau = 0.1$ s (which is a typical overall reflex delay in human balancing), a pendulum of length less than $l_{\text{crit},0} = 0.2943$ m ≈ 30 cm cannot be balanced in its upward position using a traditional PD controller.

Equation (5.166) can be written in the form

$$\dot{\mathbf{x}}(t) = \mathbf{A}(t)\mathbf{x}(t) + \mathbf{B}\mathbf{u}(t - \tau),\tag{5.168}$$

$$\mathbf{u}(t) = \mathbf{D}\mathbf{x}(t),\tag{5.169}$$

where

$$\mathbf{x}(t) = \begin{pmatrix} \varphi(t) \\ \dot{\varphi}(t) \end{pmatrix}, \quad \mathbf{A} = \begin{pmatrix} 0 & 1 \\ -(\delta + \varepsilon\cos(\omega t)) & 0 \end{pmatrix}, \quad \mathbf{B} = \begin{pmatrix} 0 \\ 1 \end{pmatrix}, \quad \mathbf{D} = \begin{pmatrix} -k_{\text{p}} & -k_{\text{d}} \end{pmatrix}.\tag{5.170}$$

Stability charts for (5.168)–(5.169) can be determined by the semi-discretization method, as shown in Section 3.3.

Figure 5.35 presents some sample stability charts in the plane $(k_{\text{p}}, k_{\text{d}})$ for a stick of length $l = 50$ cm with forcing amplitude $r_{\text{a}} = 5$ mm for different forcing frequencies f. It can be seen that the stability charts are qualitatively different from those of the time-invariant system with $\varepsilon = 0$ (see Figure 5.34). Here, the stable domains are serrated by unstable tongues caused by the parametric forcing. Equation (5.166) was also analyzed by Sheng et al. [241], where similar stability charts were presented in the plane $(k_{\text{p}}, k_{\text{d}})$, while the same system under proportional-integral (PI) control was analyzed by Sheng and Sun [240].

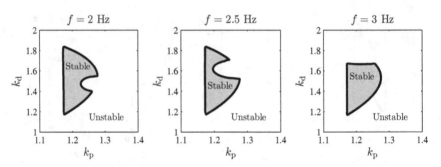

Fig. 5.35 Stability charts for (5.166) with feedback delay $\tau = 0.1$ s, stick length $l = 50$ cm, forcing amplitude $r_a = 5$ mm for different forcing frequencies f.

In order to analyze the effect of parametric forcing on the stabilization process, stability charts in the plane (k_p, k_d) are determined for a series of forcing frequencies f, forcing amplitudes r_a, and pendulum lengths l. Figure 5.36 presents a set of diagrams that describe the changes in the stability properties for different values of f, r_a, and l. Different shades of gray indicate stability boundaries for different lengths l, while the feedback delay is $\tau = 0.1$ s for all plots. These diagrams can be considered projections of the 5-dimensional stability chart in the parameter space (k_p, k_d, f, r_a, l). It can be seen that for small amplitudes and for small frequencies, stable domains arise only for the lengths $l = 40, 60, 80$, and 100 cm, of which are all larger than the critical length $l_{crit,0} = 30$ cm for the unforced system. For large amplitudes and large frequencies, stable domains arise also for the case $l = 20$ cm (see, for instance the case $r_a = 20$ mm with $f = 6$ Hz, where the boundaries for all lengths can clearly be seen).

Figure 5.37 presents the critical stick lengths for different forcing frequencies and amplitudes. The chart was determined in the following way. The frequency and amplitude were changed between 2–10 Hz and 0–50 mm, respectively, both in 100 steps providing a 100×100 grid over the plane (f, r_a). For each pair (f, r_a), the length was fixed to a starting value of $l_{start} = 10$ cm, and the stability charts were determined over a 100×100 grid in the plane (k_p, k_d) with the bounds $-0.5 \leq k_p \leq 2$ and $-0.5 \leq k_d \leq 2$ (as in Figure 5.36). Then the length of the stick was increased by $\Delta l = 2.5$ cm, and the stability charts were computed again. The process was repeated until no stable region was found within the region $-0.5 \leq k_p \leq 2$, $-0.5 \leq k_d \leq 2$. The corresponding length was considered the critical length for the pair (f, r_a).

The grayscale bar in Figure 5.37 represents the critical stick length l_{crit}. Some values are also presented in the main diagram in order to help to identify the corresponding grayscales. During the analysis of the diagram, it should be considered that the stability charts were determined numerically at discrete pairs (f, r_a) with discrete lengths l in the plane (k_p, k_d) with $-0.5 \leq k_p \leq 2$ and $-0.5 \leq k_d \leq 2$. This explains the fragmented contour lines. It can be seen that for the case, in which the forcing amplitude is zero, the critical stick length is about 36 cm, which is larger than the theoretical value $l_{crit,0} = 30$ cm for the unforced system. This is again due

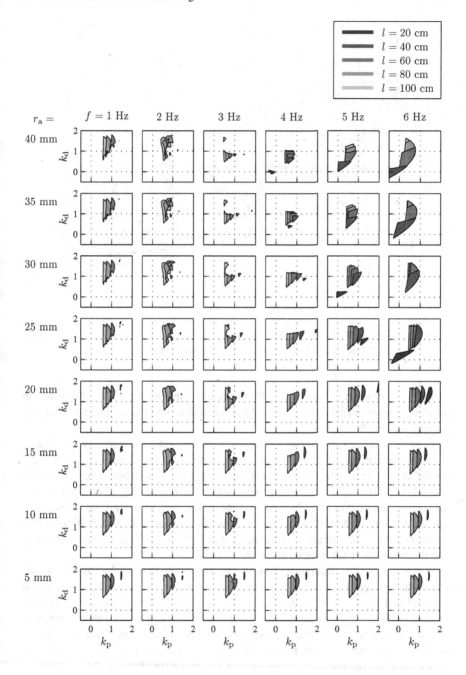

Fig. 5.36 Set of stability charts for (5.166) with $\tau = 0.1$ s for different forcing frequencies f and amplitudes r_a. Different lengths are indicated by different shades of gray. For large amplitudes and large frequencies, the stable regions appear also for length $l = 20$ cm$< l_{\mathrm{crit},0} = 30$ cm.

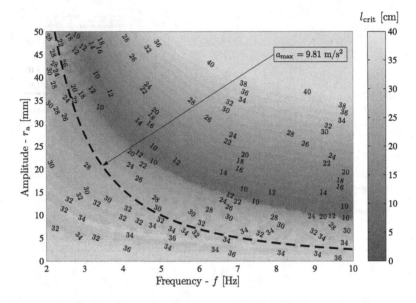

Fig. 5.37 Grayscale map for the critical length as a function of the forcing frequency and amplitude for feedback delay $\tau = 0.1$ s. Numbers denote the critical lengths. The black dashed line corresponds to the case that the maximal acceleration is equal to g.

to the discrete numerical analysis of the parameter domains: the 100×100 resolution over the parameter plane (k_p, k_d) is not fine enough to find the very narrow stability regions for lengths close to l_{crit}.

In spite of the numerical fragmentation of the diagram, the tendency can clearly be seen: the critical length of the pendulum decreases for increasing f and r_a. A radical change in the critical length can be observed at a hyperbola-like region with $l_{crit} \leq 10$ cm. This result is not surprising, since it is known that an inverted pendulum of any length can be stabilized by appropriate (i.e., high enough) forcing frequency and forcing amplitude even without feedback control. However, if the power or the maximum acceleration of the forcing is limited, then the forcing frequency and the amplitude cannot be increased arbitrarily.

A relevant issue in human stick-balancing is that the base of the stick is in contact with the fingertip, which implies that the downward acceleration cannot exceed gravitational acceleration g, i.e., $a_{max} = r_a \omega^2 \leq g = 9.81$ m/s^2 (with $\omega = 2\pi f$). This limit is indicated by the dashed line in Figure 5.37. It can be seen that parametric forcing is more effective if the forcing frequency is relatively small while the forcing amplitude is relatively large. Figure 5.37 shows that the critical length goes below 0.24 to 0.26 m even if $a_{max} \leq g$ (see around $f = 2$ to 3 Hz with $r_a = 30$ to 40 mm).

In order to confirm that a stick of length less than 30 cm can be balanced with a human reflex delay of $\tau = 0.1$ s in case of parametric forcing, a detailed stability analysis is performed for $f = 2.5$ Hz with $r_a = 30$ mm. The corresponding maxi-

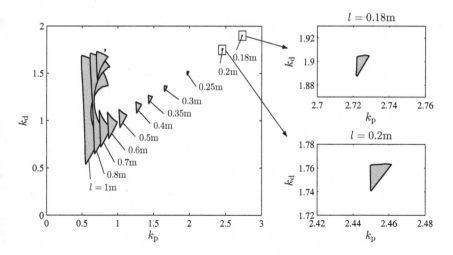

Fig. 5.38 Shift of the stability boundaries in the plane (k_p, k_d) for forcing frequency $f = 2.5\,\text{Hz}$ and forcing amplitude $r_a = 30\,\text{mm}$ for different lengths l. The maximum acceleration is $a_{\max} = 7.4\,\text{m/s}^2$. Tiny stable domains are enlarged for lengths $l = 18\,\text{cm}$ and $l = 20\,\text{cm}$ (right).

mum acceleration is $a_{\max} = 7.4\,\text{m/s}^2$, i.e, the contact between the base of the stick and the fingertip is continuously maintained. The stability charts are presented in Figure 5.38. It can be seen that very narrow stable domains do exist even for the length 18 cm. This means that stick-balancing properties can be improved by parametric forcing even if the maximum acceleration of the stick's base does not exceed gravitational acceleration g.

Appendix A
Appendix

A.1 Solution of Linear Inhomogeneous ODEs

In this appendix, the solutions for linear homogeneous and inhomogeneous ODEs are given.

Consider first the linear homogeneous ODE

$$\dot{\mathbf{y}}(t) = \mathbf{A}\mathbf{y}(t) , \tag{A.1}$$

where $\mathbf{y}(t) \in \mathbb{R}^n$ and $\mathbf{A}$ is an $n \times n$ matrix. The solution for this system associated with the initial state $\mathbf{y}(t_0) = \mathbf{y}_0$ can be given as

$$\mathbf{y}(t) = e^{\mathbf{A}t}\mathbf{y}_0 , \tag{A.2}$$

where the matrix exponential is defined by the Taylor series of the exponential function as

$$e^{\mathbf{A}t} = \exp(\mathbf{A}t) := \sum_{k=0}^{\infty} \frac{1}{k!}\mathbf{A}^k t^k , \tag{A.3}$$

with $\mathbf{A}^0 = \mathbf{I}$ being the identity matrix (see, for instance, [108, 219]). The following properties hold:

$$\frac{d}{dt}e^{\mathbf{A}t} = \mathbf{A}e^{\mathbf{A}t} = e^{\mathbf{A}t}\mathbf{A} , \qquad e^{\mathbf{A}0} = \mathbf{I} , \qquad \det\left(e^{\mathbf{A}t}\right) \neq 0 , \qquad \left(e^{\mathbf{A}t}\right)^{-1} = e^{-\mathbf{A}t} . \tag{A.4}$$

The matrix exponential $e^{\mathbf{A}t}$ can be calculated in terms of the eigenvalues and eigenvectors of $\mathbf{A}$. For instance, if $\mathbf{A}$ is a 2×2 matrix, then there exists an invertible transformation matrix $\mathbf{P}$ (whose columns consist of the generalized eigenvectors of $\mathbf{A}$) such that $\mathbf{J} = \mathbf{P}^{-1}\mathbf{A}\mathbf{P}$ has one of the following forms:

$$\mathbf{J} = \begin{pmatrix} \lambda & 0 \\ 0 & \mu \end{pmatrix} , \qquad \mathbf{J} = \begin{pmatrix} \lambda & 1 \\ 0 & \lambda \end{pmatrix} , \qquad \mathbf{J} = \begin{pmatrix} a & -b \\ b & a \end{pmatrix} , \tag{A.5}$$

where $\lambda, \mu, a, b \in \mathbb{R}$. The corresponding matrix exponentials read

$$e^{\mathbf{J}t} = \begin{pmatrix} e^{\lambda t} & 0 \\ 0 & e^{\mu t} \end{pmatrix}, \qquad e^{\mathbf{J}t} = \begin{pmatrix} e^{\lambda t} & te^{\lambda t} \\ 0 & e^{\lambda t} \end{pmatrix}, \qquad e^{\mathbf{J}t} = e^{at}\begin{pmatrix} \cos(bt) & -\sin(bt) \\ \sin(bt) & \cos(bt) \end{pmatrix}, \quad (A.6)$$

respectively. The matrix exponential $e^{\mathbf{A}t}$ can then be given by

$$e^{\mathbf{A}t} = e^{\mathbf{PJP}^{-1}t} = \mathbf{P}\,e^{\mathbf{J}t}\mathbf{P}^{-1}. \tag{A.7}$$

For $n \times n$ matrices with $n > 2$, the matrix exponential can be determined in a similar way using the Jordan form transformation of the matrix (see, for instance, [108, 219]). Matrix exponentials can be calculated by most of the numerical and symbolic software packages, as well.

Consider now the linear inhomogeneous ODE

$$\dot{\mathbf{y}}(t) = \mathbf{A}\mathbf{y}(t) + \mathbf{b}(t), \tag{A.8}$$

where $\mathbf{y}(t) \in \mathbb{R}^n$, $\mathbf{A}$ is an $n \times n$ matrix, and $\mathbf{b}(t) \in \mathbb{R}^n$ is a continuous function. The solution is determined by the method called variation of constants. The solution is searched for in the form

$$\mathbf{y}(t) = e^{\mathbf{A}t}\mathbf{g}(t), \tag{A.9}$$

where $\mathbf{g}(t) \in \mathbb{R}^n$ is a differentiable function. Every solution can be written in this form, since $e^{\mathbf{A}t}$ is invertible. Differentiation of (A.9) yields

$$\dot{\mathbf{y}}(t) = \mathbf{A}e^{\mathbf{A}t}g(t) + e^{\mathbf{A}t}\dot{\mathbf{g}}(t). \tag{A.10}$$

Substitution into (A.8) gives

$$e^{\mathbf{A}t}\dot{\mathbf{g}}(t) = \mathbf{b}(t), \tag{A.11}$$

which implies

$$\mathbf{g}(t) = \int_{t_0}^{t} e^{-\mathbf{A}s}\mathbf{b}(s)\,\mathrm{d}s + \mathbf{K}, \tag{A.12}$$

where $\mathbf{K}$ is a constant vector. The solution reads

$$\mathbf{y}(t) = e^{\mathbf{A}t}\int_{t_0}^{t} e^{-\mathbf{A}s}\mathbf{b}(s)\,\mathrm{d}s + e^{\mathbf{A}t}\mathbf{K}. \tag{A.13}$$

The initial condition $\mathbf{y}(t_0) = \mathbf{y}_0$ is satisfied if

$$\mathbf{K} = e^{-\mathbf{A}t_0}\mathbf{y}_0. \tag{A.14}$$

Thus, the solution of (A.8) for the initial condition $\mathbf{y}(t_0) = \mathbf{y}_0$ reads

$$\mathbf{y}(t) = e^{\mathbf{A}(t-t_0)}\mathbf{y}_0 + \int_{t_0}^{t} e^{\mathbf{A}(t-s)}\mathbf{b}(s)\,\mathrm{d}s. \tag{A.15}$$

This formula is called the variation of constants formula or the Duhamel–Neumann formula for linear inhomogeneous ODEs.

An Example: the Forced Oscillator

Consider the linear forced oscillator

$$\ddot{x}(t) + ax(t) = b\cos(\omega t), \tag{A.16}$$

with $a = \alpha^2 > 0$, $\omega > 0$. This system can be written in the form of (A.8) with

$$\mathbf{y}(t) = \begin{pmatrix} x(t) \\ \dot{x}(t) \end{pmatrix}, \quad \mathbf{A} = \begin{pmatrix} 0 & 1 \\ -a & 0 \end{pmatrix}, \quad \mathbf{b}(t) = \begin{pmatrix} 0 \\ b\cos(\omega t) \end{pmatrix}. \tag{A.17}$$

If $a = \alpha^2 > 0$, then the matrix exponential can be given as

$$e^{\mathbf{A}t} = \begin{pmatrix} \cos(\alpha t) & \frac{1}{\alpha}\sin(\alpha t) \\ -\alpha\sin(\alpha t) & \cos(\alpha t) \end{pmatrix}. \tag{A.18}$$

Application of the variation of constants formula (A.15) with the initial state $\mathbf{y}(0) = \mathbf{y}_0 = (x_0, v_0)^T$ gives the solution

$$
\begin{aligned}
\mathbf{y}(t) &= \begin{pmatrix} \cos(\alpha t) & \frac{1}{\alpha}\sin(\alpha t) \\ -\alpha\sin(\alpha t) & \cos(\alpha t) \end{pmatrix}\begin{pmatrix} x_0 \\ v_0 \end{pmatrix} + \int_0^t \begin{pmatrix} \frac{b}{\alpha}\sin(\alpha(t-s))\cos(\omega s) \\ b\cos(\alpha(t-s))\cos(\omega s) \end{pmatrix} ds \\
&= \begin{pmatrix} \left(x_0 + \frac{b}{\omega^2-\alpha^2}\right)\cos(\alpha t) + \frac{v_0}{\alpha}\sin(\alpha t) - \frac{b}{\omega^2-\alpha^2}\cos(\omega t) \\ v_0\cos(\alpha t) + \left(-\alpha x_0 - \frac{b\alpha}{\omega^2-\alpha^2}\right)\sin(\alpha t) + \frac{b\omega}{\omega^2-\alpha^2}\sin(\omega t) \end{pmatrix}.
\end{aligned} \tag{A.19}
$$

Alternatively, according to the theory of forced oscillators, the solution of (A.16) is searched for in the form

$$x(t) = C_1\cos(\alpha t) + C_2\sin(\alpha t) + x_{\mathrm{p}}(t), \tag{A.20}$$

where $x_{\mathrm{p}}(t)$ is the particular solution of the form

$$x_{\mathrm{p}}(t) = K\cos(\omega t) + L\sin(\omega t). \tag{A.21}$$

Substitution of (A.21) into (A.16) gives

$$K = -\frac{b}{\omega^2 - \alpha^2}, \qquad L = 0. \tag{A.22}$$

The parameters C_1 and C_2 are obtained by the substitution of the initial conditions $x(0) = x_0$, $\dot{x}(0) = v_0$, which gives

$$C_1 = x_0 + \frac{b}{\omega^2 - \alpha^2}, \qquad C_2 = \frac{v_0}{\alpha}. \tag{A.23}$$

Thus, the solution and its derivative read

$$x(t) = \left(x_0 + \frac{b}{\omega^2 - \alpha^2}\right) \cos(\alpha t) + \frac{v_0}{\alpha} \sin(\alpha t) - \frac{b}{\omega^2 - \alpha^2} \cos(\omega t) , \quad (A.24)$$

$$\dot{x}(t) = \left(-x_0\alpha - \frac{b\alpha}{\omega^2 - \alpha^2}\right) \sin(\alpha t) + v_0 \cos(\alpha t) - \frac{b\omega}{\omega^2 - \alpha^2} \cos(\omega t) , \quad (A.25)$$

which are equivalent to (A.19).

A.2 Routh–Hurwitz Criterion

The Routh–Hurwitz stability criterion is a necessary and sufficient condition for the stability of time-invariant ODEs [227, 114]. Consider an n-dimensional ODE described by the characteristic polynomial

$$p(\lambda) = a_n\lambda^n + a_{n-1}\lambda^{n-1} + \cdots + a_1\lambda + a_0 \quad (A.26)$$

with $n \geq 1$ and $a_n > 0$. The Hurwitz matrix of (A.26) reads

$$H = \begin{pmatrix} a_1 & a_0 & 0 & 0 & \ldots & 0 \\ a_3 & a_2 & a_1 & a_0 & \ldots & 0 \\ a_5 & a_4 & a_3 & a_2 & \ldots & 0 \\ \vdots & \vdots & \vdots & \vdots & & \vdots \\ a_{2n-1} & a_{2n-2} & a_{2n-3} & a_{2n-4} & \ldots & a_n \end{pmatrix} , \quad (A.27)$$

with $a_j = 0$ if $j > n$. The system is asymptotically stable (i.e., all the zeros of the polynomial (A.26) have negative real part) if and only if all the coefficients are positive, i.e., $a_j > 0$, $j = 0, 1, 2, \ldots, n$, and all the leading principal minors of H are positive, i.e.,

$$H_1 = a_1 > 0 , \quad (A.28)$$

$$H_2 = \det\begin{pmatrix} a_1 & a_0 \\ a_3 & a_2 \end{pmatrix} > 0 , \quad (A.29)$$

$$H_3 = \det\begin{pmatrix} a_1 & a_0 & 0 \\ a_3 & a_2 & a_0 \\ a_5 & a_3 & a_1 \end{pmatrix} > 0 , \quad (A.30)$$

$$\vdots$$

$$H_n = \det(H) > 0 . \quad (A.31)$$

As was shown by Farkas and Simon [82], the stability boundaries in the plane of the system parameters are determined by the conditions $a_0/a_n > 0$ and $H_n > 0$.

A.3 Matlab Code for the Delayed Mathieu Equation

The Matlab code for (4.1) is given below.

```
% Create stability charts for the delayed Mathieu equation of the form
% x"(t) + a1*x'(t) + (delta+eps*cos(2*pi/T*t))*x(t) = b0*x(t-tau)
% using the first-order semi-discretization method

clear; tic;

% PARAMETERS
a1 = 0;
eps = 2;
tau = 2*pi;             % time delay
T = 4*pi;               % principal period
stepx = 200;            % number of steps of delta
stepy = 100;            % number of steps of b0
b0_min = -1.5;          % min. value for b0
b0_max = 1.5;           % max. value for b0
delta_min = -1;         % min. value for delta
delta_max = 5;          % max. value for delta
options.disp = 0;       % options for command eigs

% SEMI-DISCRETIZATION
p = 40;                 % period resolution
h=T/p;                  % discretization step
r = floor(tau/h+1/2);   % delay resolution
eps_cos_function = @(s)eps*cos(2*pi/T*s);
for i = 1:p             % piecewise constant approx. of eps*cos(2*pi/T*t)
    eps_cos(i) = quad(eps_cos_function,(i-1)*h,i*h)/h;
end
Ai = [[0 1];[0 -a1]];
D = [1, 0];
Gi = zeros(r+2)+diag(ones(r+1,1),-1);    % constructing matrix Gi
Gi(1:3,1:2) = [zeros(2);D];

for y = 1:stepy+1                        % loop for b0
    b0 = b0_min + (y-1)*(b0_max-b0_min)/stepy;
    B = [0; b0];
    for x = 1:stepx+1                    % loop for delta
        delta = delta_min + (x-1)*(delta_max-delta_min)/stepx;
        Phi = eye(r+2);                  % constructing matrix Phi
        for i = 1:p
            Ai(2,1) = -(delta+eps_cos(i));
            Pi = expm(Ai*h);
            if det(Ai) ~= 0              % if inv(Ai) exist
                Ri0 = (inv(Ai)+1/h*(Ai^(-2)-(tau-(r-1)*h)*inv(Ai)))...
                    *(eye(2)-Pi))*B;
                Ri1 = (-inv(Ai)+1/h*(-Ai^(-2)+(tau-r*h)...
                    *inv(Ai))*(eye(2)-Pi))*B;
            else                         % if inv(Ai) does not exist
                Ri0_int = @(s)-(s-tau+(r-1)*h)/h...
                    *expm(Ai*(h-s))*B;
                Ri1_int = @(s)(s-tau+r*h)/h*expm(Ai*(h-s))*B;
```

```
                    Ri0 = quadv(Ri0_int,0,h);
                    Ri1 = quadv(Ri1_int,0,h);
                end
                Gi(1:2,1:2) = Pi;    % constructing matrix Gi
                Gi(1:2,r+2) = Ri0;
                Gi(1:2,r+1) = Ri1;
                Phi0 = Phi;          % Phi by effective matrix multiplication
                Phi(1:2,:) = Pi*Phi0(1:2,:)+[Ri1, Ri0]*Phi0((r+1):(r+2),:);
                Phi(3,:) = D*Phi0(1:2,:);
                Phi(4:r+2,:) = Phi0(3:r+1,:);
            end
            delta_m(x,y) = delta;    % matrix of delta
            b0_m(x,y) = b0;          % matrix of b0
            if r+2 > 60              % eigenvalue calculation
                try                  % for large matrices (size > ~60)
                    eig_m(x,y) = abs(eigs(Phi,1,'lm',options));    % by eigs
                catch                % in case of convergence failure
                    eig_m(x,y) = max(abs(eig(Phi)));               % by eig
                end
            else                     % for small matrices (size < ~60)
                eig_m(x,y) = max(abs(eig(Phi)));                   % by eig
            end
        end
        stepy+1-y                    % counter
end
time = toc
figure
contour(delta_m,b0_m,eig_m,[1, 1],'k')
```

A.4 Matlab Code for Multiple Delays

The Matlab code for (4.49) is given below.

```
% Create stability charts for the delayed Mathieu equation with multiple
% delays in the form
% x"(t) + a1*x'(t) + (delta+eps*cos(2*pi/T*t))*x(t)
%                                    = b01*x(t-tau1) + b02*x(t-tau2)
% using the first-order semi-discretization method

clear; tic;

% PARAMETERS
a1 = 0;
delta = 6;
eps = 6;
b01 = 1;
b02 = 1;
T = 5;              % principal period
p = 100;            % period resolution
stepx = p;          % steps of tau1
stepy = p;          % steps of tau2
```

```
tau_min = 0;            % min. value for tau1 and tau2
tau_max = 10;           % max. value for tau1 and tau2
options.disp = 0;       % options for eigs

% SEMI-DISCRETIZATION
h=T/p;                          % discretization step
r = floor(tau_max/h+1/2);       % delay resolution
eps_cos_function = @(s)eps*cos(2*pi/T*s);
for i = 1:p              % piecewise constant approx. of eps*cos(2*pi/T*t)
    eps_cos(i) = quad(eps_cos_function,(i-1)*h,i*h)/h;
end
Ai = [[0 1];[0 -a1]];
D = [1, 0];
Gi = zeros(r+2)+diag(ones(r+1,1),-1);   % constructing matrix Gi
Gi(1:3,1:2) = [zeros(2);D];

for y = 1:stepy+1                        % loop for tau2
    tau2 = tau_min + (y-1)*(tau_max-tau_min)/stepy;
    r2 = floor(tau2/h+1/2);              % delay resolution for tau2
    B1 = [0; b01];
    for x = 1:stepx+1                    % loop for tau1
        tau1 = tau_min + (x-1)*(tau_max-tau_min)/stepx;
        r1 = floor(tau1/h+1/2);          % delay resolution for tau1
        B2 = [0; b02];
        Phi = eye(r+2);                  % constructing matrix Phi
        for i = 1:p
            Ai(2,1) = -(delta+eps_cos(i)) + b01*(r1==0) + b02*(r2==0);
            Pi = expm(Ai*h);
            if det(Ai) ~= 0              % if inv(Ai) exist
                R1i0 = (inv(Ai)+1/h*(Ai^(-2)-(tau1-(r1-1)*h)*inv(Ai)))...
                    *(eye(2)-Pi))*B1 * (r1 ~= 0);
                R1i1 = (-inv(Ai)+1/h*(-Ai^(-2)+(tau1-r1*h)*inv(Ai)))...
                    *(eye(2)-Pi))*B1 * (r1 ~= 0);
                R2i0 = (inv(Ai)+1/h*(Ai^(-2)-(tau2-(r2-1)*h)*inv(Ai)))...
                    *(eye(2)-Pi))*B2 * (r2 ~= 0);
                R2i1 = (-inv(Ai)+1/h*(-Ai^(-2)+(tau2-r2*h)*inv(Ai)))...
                    *(eye(2)-Pi))*B2 * (r2 ~= 0);
            else                         % if inv(Ai) does not exist
                A1i0_int = @(s)-(s-tau1+(r1-1)*h)/h*expm(Ai*(h-s))...
                    *B1 * (r1 ~= 0);
                A1i1_int = @(s)(s-tau1+r1*h)/h*expm(Ai*(h-s))...
                    *B1 * (r1 ~= 0);
                A2i0_int = @(s)-(s-tau2+(r2-1)*h)/h*expm(Ai*(h-s))...
                    *B2 * (r2 ~= 0);
                A2i1_int = @(s)(s-tau2+r2*h)/h*expm(Ai*(h-s))...
                    *B2 * (r2 ~= 0);
                R1i0 = quadv(A1i0_int,0,h);
                R1i1 = quadv(A1i1_int,0,h);
                R2i0 = quadv(A2i0_int,0,h);
                R2i1 = quadv(A2i1_int,0,h);
            end
            Gi(1:2,:) = zeros(2,r+2);    % constructing matrix Gi
            Gi(1:2,1:2) = Pi;
            if r1 == 1                    % case if r1=1
```

```
                    Gi(1:2,1:2) = Gi(1:2,1:2) + R1i1*D;
               else
                    Gi(1:2,r1+1) = Gi(1:2,r1+1) + R1i1;
               end
               Gi(1:2,r1+2) = Gi(1:2,r1+2) + R1i0;
               if r2 == 1                        % case if r2=1
                    Gi(1:2,1:2) = Gi(1:2,1:2) + R2i1*D;
               else
                    Gi(1:2,r2+1) = Gi(1:2,r2+1) + R2i1;
               end
               Gi(1:2,r2+2) = Gi(1:2,r2+2) + R2i0;
               Phi0 = Phi;        % Phi by effective matrix multiplication
               Phi(1:2,:) = Gi(1:2,:)*Phi0;
               Phi(3,:) = D*Phi0(1:2,:);
               Phi(4:r+2,:) = Phi0(3:r+1,:);
          end
          tau1_m(x,y) = tau1;       % matrix of tau1
          tau2_m(x,y) = tau2;       % matrix of tau2
          if r+2 > 60               % eigenvalue calculation
               try                  % for large matrices (size > ~60)
                    eig_m(x,y)=abs(eigs(Phi,1,'lm',options));     % by eigs
               catch                % in case of convergence failure
                    eig_m(x,y)=max(abs(eig(Phi)));               % by eig
               end
          else                      % for small matrices (size < ~60)
               eig_m(x,y)=max(abs(eig(Phi)));                    % by eig
          end
     end
     stepy+1-y                      % counter
end
time = toc
figure
contour(tau1_m,tau2_m,eig_m,[1, 1],'k')
```

A.5 Matlab Code for Distributed Delays

The Matlab code for (4.52) is given below.

```
% Create stability charts for the delayed Mathieu equation with
% distributed delay in the form
% x"(t) + a1*x'(t) + (delta+eps*cos(2*pi/T*t))*x(t)
%                    = b0 * int_{-sigma}^{0} k(theta)*x(t+theta) dtheta
% using the first-order semi-discretization method

clear; tic;

% PARAMETERS
a1 = 0;
eps = 6*pi^2;
sigma = 1;                 % max. time delay
T = 0.5;                   % principal period
```

```
stepx = 200;              % steps of delta
stepy = 100;              % steps of b0
delta_min = -5*pi^2;      % min. value for delta
delta_max = 20*pi^2;      % max. value for delta
b0_min = -20*pi^2;        % min. value for b0
b0_max = 50*pi^2;         % max. value for b0
options.disp = 0;         % options for eigs

% SEMI-DISCRETIZATION
p = 20;                   % period resolution
h=T/p;                    % discretization step
f = ceil(sigma/h);        % resolution for the distributed delay
eps_cos_function = @(s)eps*cos(2*pi/T*s);
for i = 1:p               % piecewise constant approx. of eps*cos(2*pi/T*t)
    eps_cos(i) = quad(eps_cos_function,(i-1)*h,i*h)/h;
end
%kernel_function = @(s)1+0*s;                % different kernel functions
kernel_function = @(s)pi/2*sin(pi*s);
%kernel_function = @(s)pi/2*sin(pi*s)+13*pi/77*sin(2*pi*s);
for k = 1:f-1            % discretization of the kernel function
    kernel(k) = quad(kernel_function,-k*h,-(k-1)*h);
end
kernel(f) = quad(kernel_function,-sigma,-(f-1)*h);
r = f;                    % delay resolution
Ai = [[0 1];[0 -a1]];
D = [1, 0];
Gi = zeros(r+2)+diag(ones(r+1,1),-1);   % constructing matrix Gi
Gi(1:3,1:2) = [zeros(2);D];

for y = 1:stepy+1                        % loop for b0
    b0 = b0_min + (y-1)*(b0_max-b0_min)/stepy;
    B = [0; b0];
    for x = 1:stepx+1                    % loop for delta
        delta = delta_min + (x-1)*(delta_max-delta_min)/stepx;
        Phi = eye(r+2);                  % constructing matrix Phi
        for i = 1:p
            Ai(2,1) = -(delta+eps_cos(i));
            Pi = expm(Ai*h);
            if Ai(2,1) ~= 0              % if inv(Ai) exist
                Rik0_0 = (inv(Ai)+(inv(Ai)/2-Ai^(-2)/h)*(Pi-eye(2)));
                Rik1_0 = (-inv(Ai)+(inv(Ai)/2+Ai^(-2)/h)*(Pi-eye(2)));
            else                         % if inv(Ai) does not exist
                Rik0_int = @(s)(1/2-s/h)*expm(Ai*(h-s));
                Rik1_int = @(s)(1/2+s/h)*expm(Ai*(h-s));
                Rik0_0 = quadv(Rik0_int,0,h);
                Rik1_0 = quadv(Rik1_int,0,h);
            end
            Gi(1:2,1:2) = Pi + Rik1_0*[0; b0*kernel(1)]* D;
            for k = 1:r-1                % constructing matrix Gi
                Gi(1:2,2+k) = Rik0_0*[0; b0*kernel(k)] ...
                    + Rik1_0*[0; b0*kernel(k+1)];
            end
            Gi(1:2,2+r) = Rik0_0*[0; b0*kernel(r)];
            Phi0 = Phi;                  % Phi by effective matrix multiplication
```

```
            Phi(1:2,:) = Gi(1:2,:)*Phi0;
            Phi(3,:) = D*Phi0(1:2,:);
            Phi(4:r+2,:) = Phi0(3:r+1,:);
        end
        delta_m(x,y) = delta/pi^2;      % matrix of delta
        b0_m(x,y) = b0/pi^2;            % matrix of b0
        if r+2 > 60                     % eigenvalue calculation
            try                         % for large matrices (size > ~60)
                eig_m(x,y)=abs(eigs(Phi,1,'lm',options));    % by eigs
            catch                       % in case of convergence failure
                eig_m(x,y)=max(abs(eig(Phi)));              % by eig
            end
        else                            % for small matrices (size < ~60)
            eig_m(x,y)=max(abs(eig(Phi)));                  % by eig
        end
    end
    stepy+1-y                           % counter
end
time = toc
figure
contour(delta_m,b0_m,eig_m,[1, 1],'r')
```

A.6 Matlab Code for Time-Periodic Delays

The Matlab code for (4.79) is given below.

```
% Create stability charts for the delayed oscillator with time-varying
% delay in the form
% x"(t) + a1*x'(t) + a0*x(t) = b0*x(t-tau(t)),   tau(t) = tau(t+T) > 0
% using the first-order semi-discretization method
clear; tic;

% PARAMETERS
a1 = 0;
eps = 0.2;
tau0 = 2*pi;          % mean time delay
T = pi;               % principal period
stepx = 200;          % steps of a0
stepy = 100;          % steps of b0
a0_min = -1;          % starting value for a0
a0_max = 5;           % final value for a0
b0_min = -1.5;        % starting value for b0
b0_max = 1.5;         % final value for b0
options.disp = 0;     % options for eigs

% SEMI-DISCRETIZATION
p = 20;               % period resolution
h=T/p;                % discretization step
tau_function = @(s)tau0*(1+eps*cos(2*pi/T*s));   % time-varying delay
for i = 1:p           % piecewise constant approximation of tau(t)
    tau_i(i) = quad(tau_function,(i-1)*h,i*h)/h;
```

```matlab
        r_i(i) = floor(tau_i(i)/h+1/2);        % particular delay resolution
    end
    A = [[0 1];[0 -a1]];
    D = [1, 0];
    r = max(r_i);           % delay resolution
    Gi = zeros(r+2)+diag(ones(r+1,1),-1);      % constructing matrix Gi
    Gi(1:3,1:2) = [zeros(2);D];

    for y = 1:stepy+1                           % loop for b0
        b0 = b0_min + (y-1)*(b0_max-b0_min)/stepy;
        B = [0; b0];
        for x = 1:stepx+1                       % loop for a0
            a0 = a0_min+ (x-1)*(a0_max-a0_min)/stepx;
            Phi = eye(r+2);                     % constructing matrix Phi
            for i = 1:p
                A(2,1) = -(a0);
                P = expm(A*h);
                if A(2,1) ~= 0                  % if inv(Ai) exist
                    Ri0 = (inv(A)+1/h*(A^(-2)-(tau_i(i)-(r_i(i)-1)*h)...
                            *inv(A))*(eye(2)-P))*B;
                    Ri1 = (-inv(A)+1/h*(-A^(-2)+(tau_i(i)-r_i(i)*h)...
                            *inv(A))*(eye(2)-P))*B;
                else                            % if inv(Ai) does not exist
                    Ri0_int=@(s)-(s-tau_i(i)+(r_i(i)-1)*h)/h*expm(A*(h-s))*B;
                    Ri1_int = @(s)(s-tau_i(i)+r_i(i)*h)/h*expm(A*(h-s))*B;
                    Ri0 = quadv(Ri0_int,0,h);
                    Ri1 = quadv(Ri1_int,0,h);
                end
                Gi(1:2,:) = zeros(2,r+2);       % constructing matrix Gi
                Gi(1:2,1:2) = P;
                Gi(1:2,r_i(i)+2) = Ri0;
                Gi(1:2,r_i(i)+1) = Ri1;
                Phi0 = Phi             % Phi by effective matrix multiplication
                Phi(1:2,:) = Gi(1:2,:)*Phi0;
                Phi(3,:) = D*Phi0(1:2,:);
                Phi(4:r+2,:) = Phi0(3:r+1,:);
            end
            a0_m(x,y) = a0;                     % matrix of a0
            b0_m(x,y) = b0;                     % matrix of b0
            if r+2 > 60                         % eigenvalue calculation
                try                             % for large matrices (size > ~60)
                    eig1_m(x,y)=abs(eigs(Phi,1,'lm',options));   % by eigs
                catch                           % in case of convergence failure
                    eig1_m(x,y)=max(abs(eig(Phi)));             % by eig
                end
            else                                % for small matrices (size < ~60)
                eig1_m(x,y)=max(abs(eig(Phi)));                 % by eig
            end
        end
        stepy+1-y                               % counter
    end
    time = toc
    figure
    contour(a0_m,b0_m,eig1_m,[1, 1],'r')
```

References

1. Ackermann J (1983) Abtastregelung – Band1: Analyse und Synthese. Springer-Verlag, Berlin.
2. Alexander ME, Bowman CS, Feng Zh, Gardam M, Moghadas SM, Röst G, Wu J, Yan P (2007) Emergence of drug-resistance: implications for antiviral control of influenza pandemic. P Roy Soc B–Biol Sci 274:1675–1684.
3. Allwright JC, Astolfi A, Wong HP (2005) A note on asymptotic stabilization of linear systems by periodic, piecewise constant output feedback. Automatica 41:339–344.
4. Al-Omari JFM, Gourley SA (2005) Dynamics of a stage-structured population model incorporating a state-dependent maturation delay. Nonlinear Anal–Real 6:13–33.
5. Altintas Y (2000) Manufacturing automation: metal cutting mechanics, machine tool vibrations, and CNC design. Cambridge University Press, New York.
6. Altintas Y, Budak E (1995) Analytical prediction of stability lobes in milling. CIRP Ann–Manuf Techn 44:357–362.
7. Altintas Y, Engin S, Budak E (1999) Analytical stability prediction and design of variable pitch cutters. J Manuf Sci E–T ASME 121:173–178.
8. Anderson E, Bai Z, Bischof C, Blackford S, Demmel J, Dongarra J, Du Croz J, Greenbaum A, Hammarling S, McKenney A, Sorensen D (1999) LAPACK User's Guide. SIAM, Philadelphia.
9. Anderson RJ, Spong MW (1989) Bilateral control of teleoperators with time delay. IEEE T Automat Contr 34:494–501.
10. Andronov AA, Leontovich EA (1937) Some cases of dependence of limit cycles on a parameters. Uchenye Zapiski Gorkovskogo Universiteta 6:3–24.
11. Arnoldi WE (1951) The principle of minimized iterations in the solution of the matrix eigenvalue problem. Q Appl Math 9:1729.
12. Asada H, Slotine JJE (1986) Robot Analysis and Control. Wiley, New York.
13. Asai Y, Tasaka Y, Nomura K, Nomura T, Casadio M, Morasso P (2009) A model of postural control in quiet standing: Robust compensation of delay-induced instability using intermittent activation of feedback control. PLoS ONE 4:e6169.
14. Asl FM, Ulsoy AG (2004) Analysis of a system of linear delay differential equations. J Dyn Syst–T ASME 125:215–223.
15. Åström KJ, Wittenmark B (1984) Computer controlled systems: Theory and design. Prentice-Hall, Englewood Cliffs, NJ.
16. Bachrathy D, Insperger T, Stepan G (2009) Surface properties of the machined workpiece for helical mills. Mach Sci Technol 13:227-245.
17. Bachrathy D, Turi J, Stepan G (2011) State dependent regenerative effect in milling processes. J Comput Nonlin Dyn–T ASME, 6:041002.
18. Baker C, Ford N (1992) Stability properties of a scheme for the approximate solution of a delay-integro-differential equation, Appl Numer Math 9 357–370.

19. Balachandran B, Zhao MX (2000) A mechanics based model for study of dynamics of milling operations. Meccanica 35(2):89–109.
20. Balachandran B, Kalmar-Nagy T, Gilsinn D (2009) Delay differential equations: Recent advances and new directions. Springer Verlag, New York.
21. Baldi P, Atiya AF (1994) How delays affect neural dynamics and learning. IEEE T Neural Networ 5:612–621.
22. Bayly PV, Halley JE, Mann BP, Davies MA (2001) Stability of interrupted cutting by temporal finite element analysis. In: Proc. of the ASME 2001 Design Engineering Technical Conferences, DETC2001/VIB-21581.
23. Bayly PV, Halley JE, Mann BP, Davies MA (2003) Stability of interrupted cutting by temporal finite element analysis. J Manuf Sci E–T ASME 125:220–225.
24. Beamish D, Bhatti S, Wu JH, Jing ZJ (2008) Performance limitations from delay in human and mechanical motor control. Biol Cybern 99:43–61.
25. Bediaga I, Muñoa J, Hernández J, López de Lacalle LN (2009) An automatic spindle speed selection strategy to obtain stability in high-speed milling. Int J Mach Tool Manu 49:384–394.
26. Bellen A, Zennaro M (2003) Numerical methods for delay differential equations, Oxford University Press, Oxford.
27. Bellman R, Cooke K (1963) Differential-difference equations. Academic Press, New York.
28. Bhatt SJ, Hsu CS (1966) Stability criteria for second-order dynamical systems with time lag. J Appl Mech-T ASME 33E:113–118.
29. Blythe SP, Nisbet RM, Gurney WSC (1985) Stability switches in distributed delay models. J Math Anal Appl 109:388–396.
30. Boese FG (1989) The stability chart for the linearized Cushing equation with a discrete delay and with gamma-distributed delays. Math Anal Appl 140:510–536.
31. Boikov IV (2005) The Brockett stabilization problem. Automat Rem Contr 66:746–751.
32. Bolotin VV (1964) The dynamic stability of elastic systems. Holden-Day, San Francisco.
33. Bravo U, Altuzarra O, López de Lacalle LN, Sánchez JA, Campa FJ (2005) Stability limits of milling considering the flexibility of the workpiece and the machine. Int J Mach Tool Manu 45:1669–1680.
34. Breda D, Maset S, Vermiglio R (2005) Pseudospectral differencing methods for characteristic roots of delay differential equations. SIAM J Sci Comput 27:482–495.
35. Breda D, Maset S, Vermiglio R (2006) Pseudospectral approximation of eigenvalues of derivative operators with non-local boundary conditions. Appl Numer Math 56:318–331.
36. Brockett RW (1999) A stabilization problem. In: Blondel VD, Sontag ED, Vidyasagar M, Willems JC (eds) Open problems in mathematical systems and control theory, Springer, London.
37. Budak E (2003) An analytical design method for milling cutters with nonconstant pitch to increase stability, Part I: Theory. J Manuf Sci E–T ASME 125:29–34.
38. Budak E (2003) An analytical design method for milling cutters with nonconstant pitch to increase stability, Part II: Application. J Manuf Sci E–T ASME 125:35–38.
39. Budak E, Altintas Y (1998) Analytical prediction of chatter stability in milling, Part I: General formulation. J Dyn Syst–T ASME 120:22–30.
40. Budak E, Altintas Y (1998) Analytical prediction of chatter stability in milling, Part II: Application of the general formulation to common milling systems. J Dyn Syst–T ASME 120:31–36.
41. Burns TJ, Davies MA (1997) Nonlinear dynamics model for chip segmentation in machining. Phys Rev Lett 79:447–450.
42. Burns TJ, Davies MA (2002) On repeated adiabatic shear band formation during high-speed machining. Int J Plasticity 8:487–506.
43. Butcher EA, Bobrenkov OA (2011) On the Chebyshev spectral continuous time approximation for constant and periodic delay differential equations. Commun Nonlinear Sci 16:1541–1554.
44. Butcher EA, Ma H, Bueler E, Averina V, Szabó Zs (2004) Stability of time-periodic delay-differential equations via Chebyshev polynomials. Int J Numer Meth Eng 59:895–922.

45. Butcher EA, Bobrenkov OA, Bueler E, Nindujarla P (2009) Analysis of milling stability by the Chebyshev collocation method: algorithm and optimal stable immersion levels. J Comput Nonlin Dyn–T ASME 4:031003.
46. Butcher EA, Sari M, Bueler E, Carlson T (2009) Magnus expansion for time-periodic systems: Parameter-dependent approximations. Commun Nonlinear Sci 14:4226–4245.
47. Cabrera JL, Milton JG (2004) Stick balancing: On-off intermittency and survival times. Nonlinear Studies 11:305–317.
48. Calvetti D, Reichel L, Sorensen DC (1994) An implicitly restarted Lanczos method for large symmetric eigenvalue problems. Electron T Numer Ana 2:1–21.
49. Campbell SA, Ncube I, Wu J (2006) Multistability and stable asynchronous periodic oscillations in a multiple-delayed neural system, Physica D 214: 101119.
50. Canudas C, Siciliano B, Bastin G (1996) Theory of robot control. Springer, New York.
51. Champneys A, Fraser WB (2000) The "Indian rope trick" for a parametrically excited flexible rod: linearized analysis. P Roy Soc Lond A–Math Phy 456:553–570.
52. Cho C, Song J-B, Kim M (2008) Stable haptic display of slowly updated virtual environment with multirate wave transform. IEEE-ASME T Mech 13:566–575.
53. Cooke KL, Grossman Z (1982) Discrete delay, distributed delays and stability switches. J Math Anal Appl 86:592–627.
54. Corpus WT, Endres WJ (2000) A high-order solution for the added stability lobes in intermittent machining. In: Proc. of the ASME 2000 International Mechanical Engineering Congress & Exposition, MED-11:871–878.
55. Corpus WT, Endres WJ (2004) Added stability lobes in machining processes that exhibit periodic time variation - Part 1: An analytical solution. J Manuf Sci E–T ASME 126:467–474.
56. Corpus WT, Endres WJ (2004) Added stability lobes in machining processes that exhibit periodic time variation - Part 2: Experimental validation. J Manuf Sci E–T ASME 126:475–480.
57. Craig JJ (1986) Introduction to robotics mechanics and control. Addison-Wesley, Reading.
58. Csernák G, Pálmai Z (2007) Exploration of the chaotic phenomena induced by fast plastic deformation of metals. Int J Adv Manuf Tech 40:270–276.
59. Cullum J and Willoughby RA (1985) Lanczos algorithms for large symmetric eigenvalue computations. Birkhäuser, Boston.
60. Cushing JM (1977) Time delays in single species growth models. J Math Biol 3:257–264.
61. Davies MA, Pratt JR, Dutterer B, Burns TJ (2000) The stability of low radial immersion milling. CIRP Ann–Manuf Techn 49:37–40.
62. Davies MA, Pratt JR, Dutterer B, Burns TJ (2002) Stability prediction for low radial immersion milling. J Manuf Sci E–T ASME 124:217–225.
63. Denk R (1995) Hill's equation systems and infinite determinants. Math Nachr 175:47–60.
64. Diekmann O, van Gils SA, Lunel SMV, Walther H-O (1995) Delay equations. Springer-Verlag, New York.
65. Ding Y, Zhu LM, Zhang XJ, Ding H (2010) A full-discretization method for prediction of milling stability. Int J Mach Tool Manu 50:502–509.
66. Ding Y, Zhu LM, Zhang XJ, Ding H (2010) Second-order full-discretization method for milling stability prediction. Int J Mach Tool Manu 50:927–932.
67. Dombovari Z (2008) Private communication.
68. Dombovari Z, Wilson RE, Stepan G (2008) Estimates of the bistable region in metal cutting. P Roy Soc A–Math Phy 464:3255–3271.
69. Dombovari Z, Altintas Y, Stepan G (2010) The effect of serration on mechanics and stability of milling cutters. Int J Mach Tool Manu 50:511–520.
70. Dombovari Z, Iglesias A, Zatarain M, Insperger T (2011) Prediction of multiple dominant chatter frequencies in milling processes, Int J Mach Tool Manu 51:457–464.
71. Driver RD (1963) A two-body problem of classical electrodynamics: the one-dimensional case. Ann Phys 21:122–142.
72. Driver RD (1977) Ordinary and delay differential equations. Applied Mathematical Sciences 20, Springer-Verlag, New York.

73. Elbeyli O, Sun JQ (2004) On the semi-discretization method for feedback control design of linear systems with time delay. J Sound Vib 273:429–440.
74. Èl'sgol'c LÈ (1964) Qualitative methods in mathematical analysis. AMS, Providence.
75. Engelborghs K, Dambrine M, Roose D (2001) Limitations of a class of stabilization methods for delay systems. IEEE T Automat Contr 46:336–339.
76. Engelborghs K, Luzyanina T, Samaey G (2001) DDE-BIFTOOL v.2.00: A Matlab package for bifurcation analysis of delay differential equations. Technical Report TW-330, Department of Computer Science, K.U.Leuven, Belgium.
77. Engelborghs K, Luzyanina T, Roose D (2002) Numerical bifurcation analysis of delay differential equations using DDE-BIFTOOL. ACM T Math Software 28:1–21.
78. Engin S, Altintas Y (2001) Mechanics and dynamics of general milling cutters, Part I: helical end mills. Int J Mach Tool Manu 41:2195–2212.
79. Erneux T (2009) Applied delay differential equations, Springer, New York.
80. Faassen RPH, van de Wouw N, Nijmeijer H, Oosterling JAJ (2007) An improved tool path model including periodic delay for chatter prediction in milling. J Comput Nonlin Dyn–T ASME 2:167–179.
81. Fabiano RH, Turi J (1999) Preservation of stability under approximation for a neutral FDE. Dyn Contin Discret I 5:351–364.
82. Farkas H, Simon PL (1992) Use of the parametric representation method in revealing the root structure and Hopf bifurcation, J Math Chem 9:323–339.
83. Farkas M (1994) Periodic motions. Springer-Verlag, New York.
84. Floquet MG (1883) Équations différentielles linéaires à coefficients périodiques. Ann Sci Ecole Norm S 12:47–89.
85. Garay B (2005) A brief survey on the numerical dynamics of functional differential equations. Int J Bifurcat Chaos 15:729–742.
86. Gawthrop P (2010) Act-and-wait and intermittent control: Some comments. IEEE T Contr Syst T 18:1195–1198.
87. Gawthrop PJ, Wang L (2007) Intermittent model predictive control. P I Mech Eng I-J Sys 221:1007–1018.
88. Gawthrop PJ, Wang L (2009) Event-driven intermittent control. Int J Control 82:2235–2248.
89. Gorinevsky DM, Formalsky AM, Schneider AY (1997) Force control of robotics systems, CRC Press LLC, Boca Raton.
90. Gradišek J, Kalveram M, Insperger T, Weinert K, Stepan G, Govekar E, Grabec I (2005) On stability prediction for milling. Int J Mach Tool Manu 45:741–991.
91. Gu K, Kharitonov V, Chen J (2003) Stability of time-delay systems. Birkhäuser, Boston.
92. Guckenheimer J, Holmes P (1983) Nonlinear oscillations, dynamical systems, and bifurcations of vector fields. Springer-Verlag, New York.
93. Győri I, Hartung F, Turi J (1993) Approximation of functional differential equations with time- and state-dependent delays by equations with piecewise constant arguments. IMA Preprint Series # 1130.
94. Győri I, Hartung F, Turi J (1995) Numerical approximations for a class of differential equations with time- and state-dependent delays. Appl Math Lett 8:19–24.
95. Győri I, Hartung F, Turi J (1998) Preservation of stability in delay equations under delay perturbations. J Math Anal Appl 220:290–312.
96. Hahn W (1961) On difference differential equation with periodic coefficients. J Math Anal Appl 3:70–101.
97. Halanay A (1961) Stability theory of linear periodic systems with delay (in Russian). Rev Roum Math Pure A, 6(4):633–653.
98. Halanay A (1966) Differential equations: Stability, oscillations, time lags. Academic Press, New York.
99. Hale JK (1977) Theory of functional differential equations. Springer-Verlag, New York.
100. Hale JK, Lunel SMV (1993) Introduction to functional differential equations. Springer-Verlag, New York.
101. Hartung F, Insperger T, Stepan G, Turi J (2006) Approximate stability charts for milling processes using semi-discretization. Appl Math Comput 174:51–73.

102. Hartung F, Krisztin T, Walther H-O, Wu J (2006) Functional differential equations with state-dependent delays: theory and applications. In: Cañada A, Drábek P, Fonda A (eds) Handbook of Differential Equations, Ordinary Differential Equations, vol. 3, Elsevier, North-Holland.

103. Hassard BD (1997) Counting roots of the characteristic equation for linear delay-differential systems. J Differ Equations 136:222–235.

104. Hayes ND (1950) Roots of the transcendental equations associated with a certain differential-difference equation. J London Math Soc 25:226–232.

105. Henninger C, Eberhard P (2007) A new curve tracking algorithm for efficient computation of stability boundaries of cutting processes. J Comput Nonlin Dyn–T ASME 2:360–365.

106. Henninger C, Eberhard P (2008) Improving the computational efficiency and accuracy of the semi-discretization method for periodic delay-differential equations. Eur J Mech A-Solid 27:975–985.

107. Hill GW (1886) On the part of the motion of the lunar perigee which is a function of the mean motions of the sun and moon. Acta Math 8:1–36.

108. Hirsch MW, Smale S (1974) Differential equations, dynamical systems and linear algebra. Academic Press, Berkeley.

109. Hopf E (1942) Abzweigung einer periodischen Lösung von einer stationären Lösung eines Differentialsystems. Ber Verh Sach Akad Wiss Leipzig, Math–Nat 95:3–22.

110. Hosho T, Sakisaka N, Moriyama I, Sato M, Higashimoto A, Tokugana T, Takeyama T (1977) Study for practical application of fluctuating speed cutting for regenerative chatter control. Ann CIRP 25:175–179.

111. Hsu CS (1974) On approximating a general linear periodic system. J Math Anal Appl 45:234–251.

112. Hsu CS, Bhatt SJ (1966) Stability charts for second-order dynamical systems with time lag. J Appl Mech–T ASME 33E:119–124.

113. Hu H, Wang Z (2002) Dynamics of controlled mechanical systems with delayed feedback. Springer, Berlin.

114. Hurwitz A (1895) Über die Bedingungen unter welchen eine Gleichung nur Wurzeln mit negativen reellen Theilen besitzt. Math Ann 46:74–81.

115. Inamura T, Sata T (1974) Stability analysis of cutting under varying spindle speed. Ann CIRP 23:119–120.

116. Ince EL (1926) Ordinary differential equations. Longmans, Green and Co., London.

117. Insperger T (2002) Stability analysis of periodic delay-differential equations modeling. PhD dissertation, Budapest University of Technology and Economics, Budapest, Hungary.

118. Insperger T (2006) Act and wait concept for time-continuous control systems with feedback delay. IEEE T Contr Syst T 14:974–977.

119. Insperger T (2010) Full-discretization and semi-discretization for milling stability prediction: Some comments. Int J Mach Tool Manu 50:658–662.

120. Insperger T (2011) Stick balancing with reflex delay in case of parametric forcing. Commun Nonlinear Sci 16:2160–2168.

121. Insperger T, Stepan G (2000) Stability of high-speed milling. In: Proc. of the ASME 2000 International Mechanical Engineering Congress & Exposition, AMD-241:119–123.

122. Insperger T, Stepan G (2000) Stability of the milling process. Period Polytech Mech 44:47–57.

123. Insperger T, Stepan G (2002) Semi-discretization method for delayed systems. Int J Numer Meth Eng 55:503–518.

124. Insperger T, Stepan G (2002) Stability chart for the delayed Mathieu equation. Proc R Soc Lond A–Math Phy 458:1989–1998.

125. Insperger T, Stepan G (2004) Stability analysis of turning with periodic spindle speed modulation via semi-discretization. J Vib Control 10:1835–1855.

126. Insperger T, Stepan G (2004) Updated semi-discretization method for periodic delay-differential equations with discrete delay. Int J Numer Meth Eng 61:117–141.

127. Insperger T, Stepan G (2007) Act-and-wait control concept for discrete-time systems with feedback delay. IET Control Theory A 1:553–557.

128. Insperger T, Stepan G (2010) On the dimension reduction of systems with feedback delay by act-and-wait control. IMA J Math Control I 27, 457–473.
129. Insperger T, Mann BP, Stepan G, Bayly PV (2003) Stability of up-milling and down-milling, Part 1: Alternative analytical methods. Int J Mach Tool Manu 43:25–34.
130. Insperger T, Stepan G, Hartung F, Turi J (2005) State-dependent regenerative delay in milling processes. in: Proc. of the ASME 2005 International Design Engineering Technical Conferences, DETC2005-85282.
131. Insperger T, Stepan G, Turi J (2007) State-dependent delay in regenerative turning processes. Nonlinear Dynam 47:275–283.
132. Insperger T, Barton DAW, Stepan G (2008) Criticality of Hopf bifurcation in state-dependent delay model of turning processes. Int J Nonlin Mech 43:140–149.
133. Insperger T, Stepan G, Turi J (2008) On the higher-order semi-discretizations for periodic delayed systems. J Sound Vib 313:334–341.
134. Insperger T, Kovacs LL, Galambos P, Stepan G (2009) Act-and-wait control concept for a force control process with delayed feedback. In: Ulbrich H, Ginzinger L (eds) Motion and vibration control, Selected papers from MOVIC 2008, Springer, Garching.
135. Insperger T, Kovacs LL, Galambos P, Stepan G (2010) Increasing the accuracy of digital force control process using the act-and-wait concept. IEEE-ASME T Mech 15:291–298.
136. Insperger T, Stepan G, Turi J (2010) Delayed feedback of sampled higher derivatives. Philos T R Soc A 368:469–482.
137. Insperger T, Wohlfart R, Turi J, Stepan G (2011) Equations with advanced arguments in stick balancing models. In: Sipahi R, Vyhlidal T, Pepe P, Niculescu S-I (eds) Time Delay Systems—Methods, Applications and New Trends, Springer.
138. Iserles A (1984) Solving linear ordinary differential equations by exponentials of iterated commutators. Numer Math 45:183–199.
139. Iserles A, Nørsett SP (1999) On the solution of linear differential equation in Lie groups. Philos T R Soc Lond A 357:983–1019.
140. Ismail F, Bastami A (1986) Improving stability of slender end mills against chatter. J Eng Ind–T ASME 108:264-268.
141. Jayaram S, Kapoor SG, DeVor RE (2000) Analytical stability analysis of variable spindle speed machining. J Manuf Sci E–T ASME 122:391–397.
142. Jury EI (1962) A simplified stability criterion for linear discrete systems. P IRE 50:1493–1500.
143. Just W (2000) On the eigenvalue spectrum for time-delayed Floquet problems. Physica D 142:153–165.
144. Kabamba P (1987) Control of linear systems using generalized sampled-data hold functions. IEEE T Automat Contr 32:772–783.
145. Kalmár-Nagy T, Stepan G, Moon FC (2001) Subcritical Hopf bifurcation in the delay equation model for machine tool vibrations. Nonlinear Dynam 26:121–142.
146. Kapitsa PL (1951) Dynamic stability of a pendulum with an oscillating point of suspension (in Russian). Zh Eksper Teoret Fiz 21:588–597.
147. Khasawneh FA, Mann BP (2011) A spectral element approach for the stability of delay systems. Int J Numer Meth Eng, 86: n/a. doi: 10.1002/nme.3122.
148. Khasawneh FA, Mann BP (2011) Stability of delay integro-differential equations using a spectral element method. Math Comput Model, in press.
149. Kienzle O (1957) Spezifische Schnittkräfte bei der Metallbearbeitung. Werkstattstechnik und Maschinenbau 47:224–225.
150. Kim WS, Bejczy AK (1993) Demonstration of a high-fidelity predictive preview display technique for telerobotic servicing in space. IEEE T Robotic Autom, 9(5):698–704.
151. Kivanc EB, Budak E (2004) Structural modeling of end mills for form error and stability analysis. Int J Mach Tool Manu 44:1151–1161.
152. Kolmanovskii VB, Myshkis AD (1999) Introduction to the theory and applications of functional differential equations. Kluwer Academic Publishers, Dordrecht.
153. Kolmanovskii VB, Nosov VR (1986) Stability of functional differential equations. Academic Press, London.

154. Konishi K, Hara N (2011) Stabilization of unstable fixed points with queue-based delay feedback control, Dynam Cont Dis Ser B, in press.
155. Konishi K, Kokame H, Hara N (2011) Delayed feedback control based on the act-and-wait concept. Nonlinear Dynam 63:513–519.
156. Kovacs LL, Kövecses J, Stepan G (2008) Analysis of effects of differential gain on dynamic stability of digital force control. Int J Nonlin Mech 43:514–520.
157. Kövecses J, Kovacs LL, Stepan G (2007) Dynamics modeling and stability of robotic systems with discrete-time force control. Arch Appl Mech 77:293–299.
158. Krauskopf B, Green K (2003) Computing unstable manifolds of periodic orbits in delay differential equations. J Comput Phys 186:230–249.
159. Kuang J, Cong Y (2005) Stability of numerical methods for delay differential equations. Science Press, Beijing.
160. Kuang Y (1993) Delay differential equations with applications in population dynamics. Academic Press, New York.
161. Kudinov VA (1967) Dynamics of Tool-Lathe (in Russian), Mashinostroenie, Moscow.
162. Kuo BC (1977) Digital Control Systems. SRL Publishing Company, Champaign.
163. Kushner HJ (2008) Numerical methods for controlled stochastic delay systems. Systems & Control: Foundations & Applications. Birkhäuser, Boston.
164. Lakshmikantham V, Trigiante D (1988) Theory of difference equations, Numerical Methods and Applications. Academic Press, London.
165. Lakshmanan M, Senthilkumar DV (2010) Dynamics of nonlinear time-delay systems. Springer, Berlin.
166. Landry M, Campbell SA, Morris K, Aguilar CO (2005) Dynamics of an inverted pendulum with delayed feedback control. SIAM J Appl Dyn Syst 4:333–351.
167. Lehoucq RB, Sorensen DC (1996) Deflation techniques for an implicitly restarted Arnoldi iteration, SIAM J Matrix Anal A 17:789–821.
168. Lehoucq RB, Sorensen DC, Yang C (1998) ARPACK Users' Guide: Solution of large-scale eigenvalue problems with implicitly restarted Arnoldi methods, SIAM Publications, Philadelphia.
169. Leonov GA (2002) Brockett's problem in the theory of stability of linear differential equations. St. Petersburg Math J 13:613–628.
170. Levi M (1988) Stability of the inverted pendulum a topological explanation. SIAM Rev 30:639–644.
171. Liu L, Kalmár-Nagy T (2010) High dimensional harmonic balance analysis for second-order delay-differential equations. J Vib Control 16:1189–1208.
172. Long X-H, Balachandran B, Mann BP (2007) Dynamics of milling processes with variable time delays. Nonlinear Dynam 47:49–63.
173. Loram ID, Lakie M, Gawthrop PJ (2009) Visual control of stable and unstable loads: what is the feedback delay and extent of linear time-invariant control? J Physiol 587:1343–1365.
174. Ma H, Butcher EA (2005) Stability of elastic columns with periodic retarded follower forces. J Sound Vib 286:849–867.
175. Magnus W (1954) On the exponential solution of differential equations for a linear operator. Comm Pure Appl Math 7:649–673.
176. Malakhovski E, Mirkin L (2006) On stability of second-order quasi-polynomials with a single delay. Automatica 42:1041–1047.
177. Mancisidor I, Zatarain M, Munoa J, Dombovari Z (2011) Fixed boundaries receptance coupling substructure analysis for tool point dynamics prediction. Adv Mat Res 223:622–631.
178. Manitius AZ, Olbrot AW (1979) Finite spectrum assignment problem for systems with delays. IEEE T Automat Contr 24:541–553.
179. Mann BP, Patel BR (2010) Stability of delay equations written as state space models. J Vib Control 16:1067–1085.
180. Mann BP, Young KA, Schmitz TL, Dilley DN (2005) Simultaneous stability and surface location error predictions in milling. J Manuf Sci E–T ASME 127:446–453.
181. Mann BP, Edes BT, Easley SJ, Young KA, Ma, K (2008) Chatter vibration and surface location error prediction for helical end mills. Int J Mach Tool Manu 48:350–361.

182. Mason MT (1981) Compliance and force control for computer controlled manipulators. IEEE T Syst Man Cy 11:418–432.
183. Mathieu E (1868) Mémoire sur le mouvement vibratoire d'une membrane de forme elliptique. J Math Pure Appl 13:137–203.
184. Mennicken R (1968) On the convergence of infinite Hill-type determinants. Arch Ration Mech An 30:12–37.
185. Merdol SD, Altintas Y (2004) Mechanics and dynamics of serrated cylindrical and tapered end mills, J Manuf Sci E–T ASME 126:317–326.
186. Merdol SD, Altintas Y (2004) Multi frequency solution of chatter stability for low immersion milling. J Manuf Sci E–T ASME 126:459–466.
187. Michiels W, Niculescu S-I (2007) Stability and stabilization of time-delay systems: an eigenvalue-based approach. SIAM Publications, Philadelphia.
188. Michiels W, Engelborghs K, Vansevenant P, Roose D (2002) Continuous pole placement for delay equations. Automatica, 38:747–761.
189. Michiels W, Vyhlidal T, Zitek P (2010) Control design for time-delay systems based on quasi-direct pole placement. J Process Contr 20:337–343.
190. Milton J, Cabrera JL, Ohira T, Tajima S, Tonosaki Y, Eurich CW, Campbell SA (2009) The time-delayed inverted pendulum: Implications for human balance control. Chaos, 19:026110.
191. Milton JG, Ohira T, Cabrera JL, Fraiser RM, Gyorffy JB, Ruiz FK, Strauss MA, Balch EC, Marin RJ, Alexander JL (2009) Balancing with vibration: A prelude for "drift and act" balance control. PLoS ONE 4:e7427.
192. Milton J, Townsend JL, King MA, Ohira T (2009) Balancing with positive feedback: the case for discontinuous control. Philos T R Soc A 367:1181–1193.
193. Minis I, Yanushevsky R (1993) A new theoretical approach for the prediction of machine tool chatter in milling. J Eng Ind–T ASME 115:1–8.
194. Minorsky N (1942) Self-excited oscillations in dynamical systems possessing retarded actions. J Appl Mech–T ASME 9:65–71.
195. Mondié S, Michiels W (2003) Finite spectrum assignment of unstable time-delay systems with a safe implementation. IEEE T Automat Contr 48:2207–2212.
196. Mondié S, Dambrine M, Santos O (2002) Approximation of control laws with distributed delays: a necessary condition for stability. Kybernetika 38:541–551.
197. Moradi H, Bakhtiari-Nejad F, Movahhedy MR (2008) Tuneable vibration absorber design to suppress vibrations: An application in boring manufacturing process, J Sound Vib 318:93–108.
198. Morărescu CI, Niculescu SI, Gu KQ (2007) Stability crossing curves of shifted gamma-distributed delay systems. SIAM J Appl Dyn Syst 6:475–493.
199. Moreau L, Aeyels D (2004) Periodic output feedback stabilization of single-input single-output continuous-time systems with odd relative degree. Syst Control Lett 51:395–406.
200. Munir S, Book WJ (2002) Internet-based teleoperation using wave variables with prediction. IEEE-ASME T Mech 7:124–133.
201. Muñoa J (2007) A general model for milling stability prediction (in Spanish), PhD Thesis, Mondragon University, Mondragon, Basque Country, Spain.
202. Muñoa J, Zatarain M, Dombovari Z, Yang Y (2009) Effect of mode interaction on stability of milling processes, In: Proc. of the 12th CIRP Conference on Modeling of Machining Operations, 2:927–933.
203. Myshkis AD (1955) Lineare Differentialgleichungen mit nacheilendem Argument. Deutscher Verlag der Wissenschaften, Berlin.
204. Namachchivaya NS, Beddini R (2003) Spindle speed variation for the suppression of regenerative chatter. J Nonlinear Sci 13:265–288.
205. Nayfeh AH, Mook DT (1979) Nonlinear oscillations. John Wiley and Sons, New York.
206. Neimark Ju I (1949) D-subdivision and spaces of quasi-polynomials (in Russian). Prikl Mat Mekh 13:349–380.
207. Niculescu S-I (2001) Delay effects on stability – A robust control approach. Springer-Verlag, London.

208. Nisbet RM, Gurney WSC (1983) The systematic formulation of population models for insects with dynamically varying instar duration. Theor Popul Biol 23:114–135.
209. Ogata K (1995) Discrete-time control systems. Prentice-Hall, Englewood Cliffs.
210. Olgac N, Sipahi R (2002) An exact method for the stability analysis of time delayed LTI systems. IEEE T Automat Contr 47:793–797.
211. Olgac N, Sipahi R (2005) The cluster treatment of characteristic roots and the neutral type time-delayed systems. J Dyn Syst–T ASME 127:88–97.
212. Olgac N, Ergenc AF, Sipahi R (2005) "Delay scheduling": A new concept for stabilization in multiple delay systems. J Vib Control 11:1159–1172.
213. Orosz G, Stepan G (2006) Subcritical Hopf bifurcations in a car-following model with reaction-time delay. P Roy Soc A–Math Phy 462:2643–2670.
214. Orosz G, Krauskopf B, Wilson RE (2005) Bifurcations and multiple traffic jams in a car-following model with reaction-time delay. Physica D, 211:277–293.
215. Orosz G, Wilson RE, Szalai R, Stepan G (2009) Exciting traffic jams: Nonlinear phenomena behind traffic jam formation on highways. Phys Rev E 80:046205.
216. Orosz G, Moehlis J, Murray RM (2010) Controlling biological networks by time-delayed signals. Philos T R Soc A 368:439–454.
217. Pakdemirli M, Ulsoy AG (1997) Perturbation analysis of spindle speed variation in machine tool chatter. J Vib Control 3:261–278.
218. Pálmai Z, Csernák G (2009) Chip formation as an oscillator during the turning process. J Sound Vib 326:809–820.
219. Perko L (1996) Differential equations and dynamical systems. Springer-Verlag, New York.
220. Polushin IG, Liu PX, Lung C-H (2007) A force-reflection algorithm for improved transparency in bilateral teleoperation with communication delay. IEEE-ASME T Mech 12:361–374.
221. Pontryagin LS (1942) On the zeros of some elementary transcendental functions (in Russian). Izv Akad Nauk SSSR 6:115–134.
222. Raibert MH, Craig JJ (1981) Hybrid position/force control of manipulators, J Dyn Syst–T ASME 102:126–133.
223. Rieke F, Warland D, de Ruyter van Steveninck R, Bialek W (1997) Spikes: exploring the neural code, MIT Press, Cambridge.
224. Ronco E, Arsan T, Gawthrop PJ (1999) Open-loop intermittent feedback control: Practical continuous-time GPC. IEE P–Contr Theor Ap 146:426–434.
225. Rosenwasser E, Lampe BP (2000) Computer controlled systems: analysis and design with process-orientated models. Springer-Verlag, London.
226. Röst G, Wu J (2008) SEIR epidemiological model with varying infectivity and infinite delay. Math Biosci Eng 5:389–402.
227. Routh EJ (1877) A treatise on the stability of a given state of motion. Macmillan, London.
228. Sastry S, Kapoor SG, DeVor RE, Dullerud GE (2001) Chatter stability analysis of the variable speed face-milling process. J Manuf Sci E–T ASME 123:537–546.
229. Sastry S, Kapoor SG, DeVor RE (2002) Floquet theory based approach for stability analysis of the variable speed face-milling process. J Manuf Sci E–T ASME 124:10–17.
230. von Schlippe B, Dietrich R (1941) Shimmying of a pneumatic wheel. Lilienthal-Gesellschaft für Luftfahrtforschung, Bericht, 140:125–160, translated for the AAF in 1947 by Meyer & Company.
231. Schmitz TL, Smith KS (2009) Machining dynamics - Frequency response to improved productivity. Springer, New York.
232. Schmitz TL, Davies MA, Kennedy MD (2001) Tool point frequency response prediction for high-speed machining by RCSA. J Manuf Sci E–T ASME 123:700–707.
233. Schmitz T, Davies M, Medicus K, Snyder J (2001) Improving high-speed machining material removal rates by rapid dynamic analysis. CIRP Ann–Manuf Techn 50:263–268.
234. Segalman DJ, Butcher EA (2000) Suppression of regenerative chatter via impendance modulation. J Vib Control 6:243–256.
235. Seguy S, Dessein G, Lionel Arnaud L (2008) Surface roughness variation of thin wall milling, related to modal interactions. Int J Mach Tool Manu 48:261–274.

236. Seguy S, Insperger T, Arnaud L, Dessein G, Peigné G (2010) On the stability of high-speed milling with spindle speed variation. Int J Adv Manuf Tech 48:883–895.
237. Sellmeier V, Denkena B (2011) Stable islands in the stability chart of milling processes due to unequal tooth pitch, Int J Mach Tool Manu 51:152–164.
238. Sexton JS, Stone BJ (1978) The stability of machining with continuously varying spindle speed. Ann CIRP 27:321–326.
239. Sexton JS, Milne RD, Stone BJ (1977) A stability analysis of single point machining with varying spindle speed. Appl Math Model 1:310–318.
240. Sheng J, Sun JQ (2005) Feedback controls and optimal gain design of delayed periodic linear systems. J Vib Control 11:277–294.
241. Sheng J, Elbeyli O, Sun JQ (2004) Stability and optimal feedback controls for time-delayed linear periodic systems. AIAA J 42:908–911.
242. Shi HM, Tobias SA (1984) Theory of finite amplitude machine tool instability. Int J Mach Tool D R 24:45–69.
243. Sieber J, Krauskopf B (2004) Complex balancing motions of an inverted pendulum subject to delayed feedback control. Physica D 197:332–345.
244. Sieber J, Szalai R (2011) Characteristic matrices for linear periodic delay differential equations. SIAM J Appl Dyn Syst 10:129–147.
245. Silva GJ, Datta A, Bhattacharyya SP (2002) New results on the synthesis of PID controllers. IEEE T Automat Contr 47:241–252.
246. Sinha SC, Butcher EA (1997) Symbolic computation of fundamental solution matrices for linear time-periodic dynamical systems. J Sound Vib 206:61–85.
247. Sinha SC, Wu DH (1991) An efficient computational scheme for the analysis of periodic systems. J Sound Vib 151:91–117.
248. Sims ND (2000) Vibration absorbers for chatter suppression: a new analytical tuning methodology. J Sound Vib 301:592–607.
249. Sims ND, Mann B, Huyanan S (2008) Analytical prediction of chatter stability for variable pitch and variable helix milling tools. J Sound Vib 317:664–686.
250. Sipahi R, Olgac N (2005) Kernel and offsrping concepts for the stability robustness of multiple time delayed systems (MTDS). In Proc. of the ASME 2005 International Design Engineering Technical Conferences, DETC2005-84470.
251. Smith HL (2010) An introduction to delay differential equations with applications to the life sciences. Springer, New York.
252. Smith OJM (1958) Feedback control systems, McGraw-Hill Series in Control Systems Engineering. McGraw-Hill, New York.
253. Smith S, Tlusty J (1991) An overview of modelling and simulation of the milling process. J Eng Ind–T ASME 113:169–175.
254. Smith S, Tlusty J (1992) Stabilizing chatter by automatic spindle speed regulation. Cirp Ann–Manuf Techn 41:433–436.
255. Stepan G (1989) Retarded dynamical systems. Longman, Harlow.
256. Stepan G (1998) Delay-differential equation models for machine tool chatter. In: Moon FC (ed) Dynamics and chaos in manufacturing processes, Wiley, New York.
257. Stepan G (1998) Delay, nonlinear oscillations and shimmying wheels. In: Moon FC (ed) Applications of nonlinear and chaotic dynamics in mechanics, Kluwer Academic Publisher, Dordrecht.
258. Stepan G (2001) Vibrations of machines subjected to digital force control. Int J Solids Struct 38:2149–2159.
259. Stepan G (2009) Introduction to delay effects in brain dynamics. Philos T R Soc A 367:1059–1062.
260. Stepan G (2009) Delay effects in the human sensory system during balancing. Philos T R Soc A 367:1195–1212.
261. Stepan G, Insperger T (2006) Stability of time-periodic and delayed systems – a route to act-and-wait control. Annu Rev Control 30:159–168.
262. Stepan G, Kalmár-Nagy T (1997) Nonlinear regenerative machine tool vibrations. In: Proc. of the 1997 ASME Design Engineering Technical Conferences, DETC97/VIB-4021.

263. Stepan G, Kollar L (2000) Balancing with reflex delay. Math Comput Model 31:199–205.
264. Stepan G, Steven A, Maunder L (1990) Design principles of digitally controlled robots. Mech Mach Theory 25:515–527.
265. Stephenson A (1908) On a new type of dynamical stability. Memoirs and Proceedings of the Manchester Literary and Philosophical Society 52:1–10.
266. Strutt JW (Lord Rayleigh) (1887) On the maintenance of vibrations by forces of double frequency, and on the propagation of waves through a medium endowed with a periodic structure. Philosophical Magazine and Journal of Science 24:145–159.
267. Sun JQ (2009) A method of continuous time approximation of delayed dynamical systems. Commun Nonlinear Sci 14:998–1007.
268. Sun JQ, Song B (2009) Control studies of time-delayed dynamical systems with the method of continuous time approximation. Commun Nonlinear Sci 14:3933–3944.
269. Szalai R, Stepan G (2003) Stability boundaries of high-speed milling corresponding to period doubling are essentially closed curves. In: Proc. of the ASME 2003 International Mechanical Engineering Conference and Exposition, IMECE2003-42122.
270. Szalai R, Stepan G (2006) Lobes and lenses in the stability chart of interrupted turning. J Comput Nonlin Dyn–T ASME 1:205–211.
271. Szalai R, Stepan G (2010) Period doubling bifurcation and center manifold reduction in a time-periodic and time-delayed model of machining. J Vib Control 16:1169–1187.
272. Szalai R, Stepan G, Hogan SJ (2006) Continuation of bifurcations in periodic delay-differential equations using characteristic matrices. SIAM J Sci Comput 28:1301–1317.
273. Takacs D, Stepan G, Hogan SJ (2008) Isolated large amplitude periodic motions of towed rigid wheels. Nonlinear Dynam 52:27–34.
274. Takacs D, Orosz G, Stepan G (2009) Delay effects in shimmy dynamics of wheels with stretched-string like tyres. Eur J Mech A-Solid 28:516–525.
275. Takemura T, Kitamura T, Hoshi T, Okushima K (1974) Active suppression of chatter by programmed variation of spindle speed. Ann CIRP 23:121–122.
276. Tarng YS, Kao JY, Lee EC (2000) Chatter suppressions in turning operations with a tuned vibration absorber. J Mater Process Tech 105:55–60.
277. Taylor FW (1907) On the art of cutting metals. Transactions of ASME, 28:31–350.
278. Tlusty J (1986) Dynamics of high-speed milling. J Eng Ind–T ASME 108:59–67.
279. Tlusty J (2000) Manufacturing processes and equipment, Prentice Hall, New Jersey.
280. Tlusty J, Polacek A, Danek C, Spacek J (1962) Selbsterregte Schwingungen an Werkzeugmaschinen. VEB Verlag Technik, Berlin.
281. Tobias SA (1965) Machine tool vibration. Blackie, London.
282. Tobias SA, Fishwick, W (1958) Theory of regenerative machine tool chatter. The Engineer, Feb. 199–203, 238–239.
283. Tsao TC, McCarthy MW, Kapoor SG (1993) A new approach to stability analysis of variable speed machining systems. Int J Mach Tool Manu 33:791–808.
284. Tsypkin YaZ (1946) The systems with delayed feedback. Avtomatika i Telemekhanika 7:107–129.
285. Turner S, Merdol D, Altintas Y, Ridgway K (2007) Modelling of the stability of variable helix end mills. Int J Mach Tool Manu 47:1410–1416.
286. van der Pol F, Strutt MJO (1928) On the stability of the solutions of Mathieu's equation. Philosophical Magazine and Journal of Science 5:18–38.
287. Vyhlídal T, Zítek P (2009) Mapping based algorithm for large-scale computation of quasipolynomial zeros. IEEE T Automat Contr 54:171–177.
288. Volterra V (1928) Sur la théorie mathématique des phénomènes héréditaires. J Math Pure Appl 7:149–192.
289. Wahi P (2005) A study of delay Differential equations with applications to machine tool vibrations, PhD thesis, Indian Institute of Science, Bangalore, India.
290. Wahi P, Chatterjee A (2005) Galerkin projections for delay differential equations. J Dyn Syst–T ASME 127:80–87.
291. Wahi P, Chatterjee A (2008) Self-interrupted regenerative metal cutting in turning. Int J Nonlin Mech 43:111–123.

292. Walther HO (2002) Stable periodic motion of a system with state-dependent delay. Differential and Integral Equations 15:923–944.
293. Wan M, Zhang WH, Dang JW, Yang Y (2010) A unified stability prediction method for milling process with multiple delays. Int J Mach Tool Manu 50:29–41.
294. Wang Q-G, Lee TH, Tan KK (1998) Finite spectrum assignment for time-delay systems. In: Lecture Notes in Control and Information Sciences, 239, Springer.
295. Whitney DE (1977) Force feedback control of manipulator fine motion. J Dyn Syst–T ASME 98:91–97.
296. Wu D, Chen K, Wang X (2009) An investigation of practical application of variable spindle speed machining to noncircular turning process. Int J Adv Manuf Tech 44:10941105.
297. Yang Y, Munoa J, Altintas Y (2010) Optimization of multiple tuned mass dampers to suppress machine tool chatter. Int J Mach Tool Manu 50: 834–842.
298. Yi S, Nelson PW, Ulsoy AG (2010) Eigenvalue assignment via the Lambert W function for control for time-delay systems. J Vib Control 16:961–982.
299. Yi S, Nelson PW, Ulsoy, AG (2010) Time-delay systems: Analysis and control using the Lambert W function. World Scientific, New Jersey.
300. Yilmaz A, AL-Regib E, Ni J (2002) Machine tool chatter suppression by multi-level random spindle speed variation. J Manuf Sci E–T ASME 124:208–216.
301. Yusoff AR, Sims ND (2011) Optimisation of variable helix tool geometry for regenerative chatter mitigation. Int J Mach Tool Manu 51:133–141.
302. Zatarain M (2008) Private communication.
303. Zatarain M, Munoa J, Peigne G , Insperger T (2006) Analysis of the influence of mill helix angle on chatter stability. CIRP Ann–Manuf Techn 55:365–368.
304. Zatarain M, Bediaga I, Munoa J, Lizarralde R (2008) Stability of milling processes with continuous spindle speed variation: Analysis in the frequency and time domains, and experimental correlation. CIRP Ann–Manuf Techn 57:379–384.
305. Zhao MX, Balachandran B (2001) Dynamics and stability of milling process. Int J Solids Struct 38:2233–2248.
306. Zhong Q-C (2006) Robust control of time-delay systems. Springer-Velag, London.